Science, Technology,
and
Social Change

Science, Technology, and Social Change

Steven Yearley

*Queen's University,
Belfast*

London
UNWIN HYMAN
Boston   Sydney   Wellington

Published by the Academic Division of
Unwin Hyman Ltd
15/17 Broadwick Street, London W1V 1FP

Unwin Hyman Inc.,
8 Winchester Place, Winchester, Mass. 01890, USA

Allen & Unwin (Australia) Ltd,
8 Napier Street, North Sydney, NSW 2060, Australia

Allen & Unwin (New Zealand) Ltd in association with
the Port Nicholson Press Ltd,
60 Cambridge Terrace, Wellington, New Zealand

First published in 1988

British Library Cataloguing in Publication Data

Yearley, Steven
    Science, technology, and social change.
    1. Scientific knowledge – Sociological
    perspectives
I. Title
306'.45

ISBN 0–04–301258–2
ISBN 0–04–301259–0 Pbk

Library of Congress Cataloging-in-Publication Data

Yearley, Steven
    Science, technology, and social change.
    Bibliography: p.
    Includes index.
    1. Science—Social aspects. 2. Technology—Social
aspects. I. Title

Q175.5.Y43 1988   303.4'83   88–5613
ISBN 0–04–301258–2 (alk. paper)
ISBN 0–04–301259–0 (pbk. : alk. paper)

Typeset in 10 on 11 point Bembo
and printed in Great Britain by
Billing and Sons Ltd, London and Worcester

Preface

As writers on the sociology of science tirelessly point out, there has been a great change in the last fifteen years in our understanding of science. It is only during this period that people studying the role of science in society have begun to make 'anthropological' excursions into the world of the laboratory. Until recently the principal products generated by scientists, their theories, reported findings and experiments, were studiously avoided by most sociologists of science. All this has now changed. Researchers have made long field trips to laboratories or have closely questioned scientists about why they believe the things they do and how they evaluate experiments and observations. As a result we now understand a good deal about the life world of science and have a more sophisticated view of the characteristics of scientific knowledge.

But this increase in understanding has been achieved at the cost of specialization. There is now an area of study within sociology specifically dedicated to the sociology of scientific knowledge. The aim of this book partly runs counter to this specializing trend. My aim is to provide a review of the principal ways in which science and technology currently relate to social change, both in the West and in the underdeveloped world. A second ambition is to show that the view of science which has emerged from the recent, detailed studies of the scientific life world supplies a useful general basis on which to build sociological analyses of science, technology and social change.

I should like to take this opportunity of thanking two friends: Michael Mulkay for his very instructive comments on the argument and structure of the book and Steve Bruce for his careful reading of the text and for many helpful recommendations. I apologize to both of them for not making better use of their suggestions. The book and its shortcomings remain my responsibility alone.

Contents

Introduction:
What Has Sociology to Say About Science and Development?

Introduction

In the past three decades the occupants of the industrialized parts of the world have become increasingly convinced of the importance of science and technology in social change. Although in the 1950s they knew they had never had it so good in terms of material wealth, welfare benefits and employment opportunities, it was also true that they had never before reaped the benefits of science and technology in the same way. Dependence on the domestic use of distinctly technological artefacts such as the television and refrigerator became widespread and brought the link between society and technology to public attention. The impact of science and technology on social change had become an obvious fact. The close connection between social development and technical ability has become an even more acute issue within the last decade, particularly in Britain but also in the long–industrialized countries of continental Europe and North America. Governments in these countries have lately come to insist on the economic benefits which can and must be derived from science and technology and have adopted an increasingly utilitarian attitude to the support and commissioning of research.

Yet this conviction of the importance of science and technology for development is as imprecise as it is widespread. The intricate details of scientific knowledge remain the preserve of a highly trained élite. Under these conditions, a general vagueness about the processes involved in the generation of scientific and technical

1

knowledge is readily understood. Even the more specialized agencies which have tried to comprehend the impact of technology commonly treat it as a factor existing entirely outside society; that is, as an exogenous influence. Thus, while economists' models can easily cope with changes in the availability or prices of raw materials, for example, the introduction of a new technology is commonly regarded as a non–economic change. It tends to be handled as though it occurred outside the economic system altogether. Sociologists' responses have generally been similar. Great reliance is still placed on the theories of social change associated with Weber, Marx and Durkheim which stress factors such as religious belief and work motivation, political struggle and the contrast between pre–industrial and modern forms of social cohesion. This is not to say that there is no consideration of technical change in the work of these theorists; rather it is to imply that sociologists have looked towards social causes of change and have not incorporated technical change in the heart of their models. Even policy–makers have been unsure about the mechanisms linking science, technology and society. Their attitudes have fluctuated greatly; for a considerable period in the 1960s it seems to have been felt that any money spent on science and technology was automatically well spent because of the expected spin–offs. Scientists, encouraged by the generosity with which they had been treated, saw the costs of research soar. A counter–reaction set in among planners and politicians which has only been unevenly relaxed as governments have sought to harness science to the national economic interest. It has been hard to know how to treat, even how to classify, this goose which occasionally lays such conspicuously golden eggs (Rip, 1982).

If the workings of science and technology at this level still retain a considerable mystique, there is another level at which the concept of science, as opposed to any specific scientific idea, has attained general currency. Science has come to have the status of exemplary knowledge. The widespread recognition of its authority is attested to in the advertising of consumer goods. Zanussi products, for example, are promoted with the caption 'The appliance of science', while Creda cookers are designated 'Science for womankind'. Scientific authority is clearly expected to command public respect.

At earlier times in Western history other types of knowledge, such as logical deduction or legal principle, were pre–eminent. Science is held to differ from these forms of knowledge because of its empirical foundation. The earlier exemplars derived their authority from their certainty; they were true by definition. These

2

bodies of knowledge were rather static and tended to be conservative. Scientific knowledge is not static; scientific theories are subject to change, and scientists aim to offer new understandings of the surrounding world.

More common still has been the preference for religion as the model of certainty. Basalla (1967, p. 617) cites the following poetic extract as displaying the Confucian response to the claims of science:

> With a microscope you see the surface of things.
> It magnifies them but does not show you reality.

Religious truth has the appeal of offering direct and certain knowledge, and the conflict between religious belief and the claims of science continues for many individuals in industrialized societies today.

But the model of knowledge with most influence in this century is the scientific one. That this influential position is not simply a recent development is indicated by Cannon in her study of High Church Anglicans in Oxford in the early Victorian period. She reports (1978, p. 12) that even churchmen who were opposed to science and its spread within the universities were already at pains to show that their knowledge was every bit as good as science. Arguing for the value of religious learning, churchmen such as John Newman claimed that 'Religious doctrine is knowledge, in as full a sense as Newton's doctrine is knowledge' (Cannon, 1978, p. 13). Even in their opposition they were tacitly acknowledging the authority of the scientific way of knowing. While this regard for science was far from uniformly spread through society in the nineteenth century, science's authority was already substantial. Even the monarchy subjected itself to it, as the following story, related by the biographer of Lyon Playfair (the first British scientist of note to become a government minister), indicates (Reid, 1899, p. 201):[1]

> whilst the Prince [of Wales] was living in Edinburgh as Playfair's pupil ... an incident occurred ... The Prince and Playfair were standing near a cauldron containing lead which was boiling at white heat.
> 'Has your Royal Highness any faith in science?' said Playfair.
> 'Certainly,' replied the Prince.
> Playfair then carefully washed the Prince's hand with ammonia to get rid of any grease that might be on it.
> 'Will you now place your hand in this boiling metal, and

3

ladle out a portion of it?' he said to his distinguished pupil.
'Do you tell me to do this?' asked the Prince.
'I do,' replied Playfair. The Prince instantly put his hand
into the cauldron, and ladled out some of the boiling lead
without sustaining any injury.

Greater adherence by world leaders to the advice of scientists
could not be looked for, even today.

This aspect of science, its authority, has been of interest to
sociologists since Weber's celebrated analysis of the bases of social
authority. He proposed that there exist three forms of authority:
the traditional, the charismatic and the legal–rational. For Weber,
science was a principal instance of the legal–rational form of
authority since claims to expertise founded on scientific understanding could be demonstrated from first principles. Such claims
to authority could be backed up with proof or demonstration.
More recent attention has focused on drawbacks in this assumption
since the promise that one *could* justify statements from first
principles often becomes substituted for the actual demonstration.
The judgements of 'experts' are often accepted because of the
position occupied by such persons (as doctors, for example) or
on account of some token of their expertise (such as a degree
certificate). The belief in the authority of science, scientism, can
readily lead into a form of traditional authority where the experts
are no longer routinely held accountable.

One area where this has provoked particular interest is in
the political sphere as scientific advisers have come to occu-
py consultative positions. Scientists may well be employed to
pass judgements about the safety of particular medical practices
or of power–generating installations. There are two forms of
anxiety which may arise here. On the one hand, given the
authority of experts' judgements, hasty, partial, or even corrupt
recommendations may be allowed to pass just because of the
person making them. More subtle is the possible outcome, often
described as the 'scientization of politics' (Habermas, 1971, p. 62),
whereby all judgements come to be regarded as technical matters.
Proponents of the concept of scientization fear that, with the spread
of technical decision–making, the impression will be fostered that
all political problems are susceptible of technical resolution and
that, in every case, there is one correct solution. The anxiety is
that such a tendency would act to pre–empt opposing viewpoints.
Scientization is thus seen as potentially anti–democratic since it
can be used to legitimate authoritarian political movements. This

implies not so much a manipulation of scientific authority as an over–extension of scientism into areas where it is not appropriate. We will return to this topic in Chapter 5 in the context of military expenditure and the formulation of defence policy.

Science, Technology, and Modernization

So far the emphasis has been on changes in the productive forces and politics of the industrialized world, but the implications are just as significant for the 'developing' or 'underdeveloped' countries. In one of the only large–scale statistical comparisons of national scientific activity, Frame (1979) has shown how closely scientific output – in terms of the number of published scientific papers and the amount spent on science – correlates with the state of development of countries as measured in economic terms. This finding need not be all that surprising, since one would anticipate that some such correlation would hold for virtually all forms of cultural activity, such as the provision of symphony orchestras. Yet the importance which we attach to science and technology in economic development indicates that the opportunities of the world's poorer countries are adversely affected by their small complement of scientific personnel. A sociological examination of the social role played by science and technology will need to study their contribution to social change in the underdeveloped world.

Such a study should be placed in the context of broader academic conceptions of underdevelopment. The concern among Western academic analysts with the modernization of Third World or undeveloped countries can be dated to the period immediately after the Second World War. These countries, newly independent after the political shake-up of the war period, presented a human and strategic problem. In the context of trying to decide how to go about helping these societies to emulate the successful (i.e. Western) nations, modernization theory was produced. Although it appeared in a number of forms, the fundamental assumptions of modernization theory were notably consistent, and it generated a particular vocabulary for designating Third World countries. Typically, advocates of modernization theory regarded the un-developed countries as lacking in some feature which had been present in the (European) nations which successfully modernized. Thus, they might lack the entrepreneurial drive associated with

Weber's Protestant Ethic or its functional equivalent. Once the deficiency could be identified, assistance would be targeted at this area, and development could commence. Such development, it was assumed, would proceed along the path pioneered by the industrialized West.

The notion that there was a particular path to be followed was fully represented in Rostow's famous taxonomy. This detailed five stages through which societies had to pass in order to reach the developed condition represented by the United States. Social development is treated as beginning with traditional societies in which landownership is the central economic principle and in which innovation and change are infrequent. This period is succeeded by the 'preconditions for take–off'; this stage witnesses institutional and conceptual changes which disrupt the previously stable pattern. The period of take–off which follows represents the stage in which 'Growth becomes [society's] normal condition' (Rostow, 1971, p. 7). The drive to maturity sees the consolidation of industrial production, the extensive use of increasingly advanced technology and the expansion of the manufacturing sector away from the original processes and products. The age of mass consumption succeeds maturity and depends upon the ability to generate a considerable manufacturing surplus. New goods are available for the population, and there is increased emphasis on service industries and on leisure. It should be noted that Rostow explicitly based these stages on the progression of England from the time of the Industrial Revolution to early in the twentieth century and thereafter on the United States experience. He openly assumes that the developed countries represent a precise model of the future of other countries.

This assumption leads to a variety of problems which I shall turn to shortly, but contemporary with these theoretical formulations were practical developments of great significance. The transfer of technology to poorer countries appeared to offer a practical route to modernity, allowing them to increase their output so as to trade and even compete with the industrialized nations. Much of the available technology was supplied through the activities of multinational enterprises. These companies were eager to establish plants and manufacturing facilities in areas where raw materials were readily available, where labour was inexpensive and where there might be access to new markets. Such transfers were usually made with conditions attached to prevent the newly acquired technology from being used to compete directly with the provider. The recipient countries also lacked the scientific personnel to develop the technical apparatus much further and thus were

6

reliant on new instalments of technology. Worse still from their point of view, this route to modernity was one being sought by a large number of countries. They were thus often competing as sites for investment and were effectively underbidding each other in a Dutch auction for advanced companies' attention. The enterprises transferring the technology were able to place more and more conditions on its acceptance. In this way the recipient countries were led into a form of subordination or dependency by the very measures which had been thought to offer them a chance of independent development.

Such a trend towards dependency has figured largely in the responses which modernization theory has excited. It has been extensively criticized, and its fundamental assumptions have been attacked. Modernization theory implicitly assumes that all countries start from a state of non–development (Rostow's traditional society) which means that they are reasonably comparable. They may be differentiated by climate or physical geography, but the state of social organization is similar. This echoes Durkheim's characterization of society as proceeding from one pervasive form of social organization – mechanical solidarity – to a more advanced state – organic solidarity. Only on these grounds could it be acceptable for Rostow, for example, to model the development of Latin American or South–East Asian countries on the economic history of England. But the countries do not start off on comparable terms. When England industrialized it was largely a pioneer. At the time of continental European industrialization, through the nineteenth century, there were still great areas of the globe where no one was engaged in industrial manufacturing and the relevant raw materials were not in demand. At the present, countries which are seeking to follow the Western path are competing for market opportunities against well established and vastly productive industrial powers.

The situation of these countries would be bad enough were they facing only this competition with stronger economies. In fact, in some areas (including food production and clothing manufacture) they face 'unfair' competition. Many underdeveloped countries have a natural advantage in producing foodstuffs arising from their geographical location. Yet food imports into the European Community, for example, are restricted so as to maintain the viability of European production. Similarly, many steps in clothing manufacture, such as the stitching up of shirts, tend to be labour intensive and here underdeveloped countries would be expected to have a price advantage arising from their lower wage rates. Yet

7

there exists the Multifibre Arrangement (MFA) which prevents completely open competition with the industrialized nations' garment makers. The MFA was renegotiated in July and August 1986 and continues to place limits on the trading activities of Third World producers (*The Times*, 1 August 1986, p. 18). European governments have good reasons for acting in these ways. For example, it can reasonably be maintained that for strategic reasons a domestic source of foodstuffs is vital. But these restrictions dictate that underdeveloped economies face a very steep uphill struggle. Evidently the currently industrialized countries did not face such limitations during their 'modernizations'. On the contrary, they were able to exert control over their colonies or impose terms of trade which acted in just the opposite way to these restrictions. They were able to dictate which unprocessed goods they imported and which manufactured goods they exported. In the course of industrialization they used the resources of other countries for their own benefit. The price of their development was often the *under*development of their colonies.

The fundamental analogy of modernization theory is thus erroneous in at least four respects. Underdeveloped countries are not in the same open environment as the non–developed nations of Europe were; generally speaking they do not have other countries as colonies on which to prey; underdeveloped countries are actually actively disadvantaged in some areas of trade where they might otherwise be confident of success; and they do not start off from a state of non–development: in the past they may well have been colonies which were forcibly underdeveloped. For these reasons the terminology used to refer to 'Third World' countries is particularly sensitive. The term used to denote these countries is much more than a matter of convenience for it embodies a range of assumptions about the kind of social change which they are experiencing. Hereafter they will generally be described as underdeveloped.

Developmental Patterns in Science and Society

The study of science and the study of development show an interesting and important similarity. Until very recently both have been studied primarily with the benefit of hindsight. In the case of social development, as we have just seen, the problem has often been posed as: how can the poorer countries get on the same track as the successful countries and emulate their success? The direction of progress has seemed obvious; the only question is how best to move in that direction. With regard to science the situation has appeared equally clear cut. Science is commonly regarded as an edifice built upon the collection of more and more data. Theories come and go as our factual knowledge increases. The only sociologically or historically significant question concerns the conditions which facilitate a proper attitude to the revelation of this truth. Analysts have felt free to use contemporary knowledge to assess the achievements of past scientists, and the scientific views of former ages have been evaluated in current terms. In this way it has often seemed peculiar that earlier scientists could have been so wrong. Moreover, the composition of science itself has been treated in this ahistorical way. Scholars have often looked for the origins of science or the factors which gave rise to science as though once discovered science would have an impulsion of its own.

In both these cases the type of explanation invoked is a form of idealism. Amongst philosophers, idealism generally refers to the claim that ideas themselves have some kind of transcendent reality. The sense of idealism being used here is slightly different and is readily understood in relation to the case of science. If we accept that science is an attempt to capture the external world in the form of concepts, then science can be pictured as successively drawing closer and closer to *the* correct image of the world's constituents. Steps in the right direction would not be the same as movements away from the truth. Progress would be natural and lack of progress problematical. Such a conception of the advancement of science is idealistic in two senses. First, it indicates that ideas and the deductions or implications which follow from them are the leading causal influence on changes in people's beliefs. It secondly implies that science is moving towards a final state of completion. In many respects this is a natural and intuitively satisfying way of thinking about science but, as will be seen shortly, it has led to some major faults in analyses of science. Before turning to the shortcomings of this approach it will be helpful to examine idealism in relation to social development.

Idealism in development is illustrated by the case of modernization theory where there is a pervasive assumption that we know what the modern or developed condition is like. Progress is then understood as any move towards this state. In this case there is little defence for the idealistic viewpoint since there is no standard outside of past and existing societies to which societal development can approximate. Of course we may derive schemes from political philosophy and measure societies up against them; we may, for example, offer an image of the best society as being liberal and individualist or as collectivist and socialist. Such a procedure is fraught with problems, however, since we can have no guarantee that any imagined societies will actually be viable. Furthermore, attempts to chart modernization towards, say, a liberal democracy are often complicated because attractive goals appear to be in short term conflict. We may encourage movements towards political democracy but also know that the likely victors in an election will not guarantee the human rights of all their citizens.

Overall we may agree that it is worthwhile modelling one society on the attractive features of another. Yet we have to acknowledge that all social arrangements are human constructions. Even if we appeal to apparently timeless values such as democracy, the form which democracy will take in a large industrialized nation resembles only slightly the ancient democracies of the Greek city states. And even within the relative cultural uniformity of Europe the existing democracies have very different electoral and party systems. There is no complete, pre–existing model of the future society to which societies can aspire. Idealism in social development is thus untenable. It is considerably more difficult to reach the same conclusion for science, but it is to this important issue that we turn in the next section.

The Critique of Idealist Theories of Science

In criticizing the assumption that there exists an ideal developmental trajectory to which countries can be compared, the point was made that all societies are human constructions. Social arrangements are humanly devised or result from the unforeseen consequences of human activity. Even general principles such as democratic rights are inventions. They are proposed by particular individuals and are subsequently taken up, adapted and re–used by other persons. This aspect of human conventions leads to them commonly being described as social constructions. The constructionist view is, in most respects, the opposite of the idealist interpretation. In the case of social development, the constructionist claim would be that politicians and analysts in the industrialized world have come to regard their past as the natural and universal state and have constructed a theory about developmental paths which implies that Western social development is entirely natural.

10

If it can be shown that a process such as social modernization has occurred through, for example, the exploitation of other countries' raw materials and labour then it is much harder to maintain that development is the unfolding of some scheme or plan. Idealism and 'constructedness' are generally incompatible.

What then of idealism in analyses of scientific progress? Two principal standpoints in the sociological analysis of science can be identified, both of which are opposed to idealism: a social construction view and a political economy view. To begin with the political economy view, its proponents argue that the development of scientific and technical knowledge is recurrently shaped by commercial and political priorities. Such priorities can clearly specify what sorts of science are pursued: whether nuclear physics or organic chemistry receives most funding and staff. Whole potential branches of science may not receive any attention if they offer no commercial or other rewards to researchers. An instance here would be the so-called orphan drugs which may benefit only very few sufferers from rare diseases and would not be commercially viable. A similar sort of influence has been identified by Slack (1972), who argues that the discipline of chemistry has concentrated on relatively rapid, unidirectional reactions because of their significance to the chemical industry. This has affected the whole slant of chemistry and the chemical conception of the natural world. While chemical processes operating very slowly over vast geological periods may be of great potential importance (say, in the mechanism of rock formation or in the origination of biochemically significant molecules), attention has generally not been focused on them. Political economic effects can be on a large, obvious scale, such as the fashionability of cancer research or the plentiful funding for work on the United States 'strategic defence initiative'; they may also be much more subtle, as Slack's argument about the self–image of chemistry indicates.

In these cases the central argument is that scientific knowledge is being filtered and selected by commercial and political considerations. It may even be that potentially available knowledge is being withheld for political economic reasons. We may agree about the practical importance of such claims but doubt their connection to idealistic views of science. Political economy contradicts idealist interpretations of science because it shows how economic and socio–political considerations influence the way that scientific knowledge develops. It shows that there is no single, pre–set route along which knowledge unfolds. The path of science is inextricably bound up with its social and economic context.

11

When this perspective is systematically applied to the history of science, even if only to the period since the eighteenth century, it is easy to see that the whole scientific enterprise could have been vastly different. We might now know very different things; we might think very differently. The growth of knowledge could be regarded idealistically only if there were some guarantee that there was just one path along which science could advance. Instead there are many conceivable routes which scientific knowledge could have taken, and the one which we experience has been specified for us by economic and political factors. In this way, what we commonly regard as the neutral, value–free state of science can be regarded as a construction, or at best a permutation, based on a great variety of political, commercial and personal considerations. On this view, sociology clearly has much to say about why we have the science we do.

The second approach starts out from the proposal (to be enlarged on in Chapter 1) that scientific knowledge is not a simple reflection of natural reality nor does it derive straightforwardly from experiments and observations. Of course, scientific knowledge is empirically based. It depends on observations and experiments but it also goes beyond them. Science is thus a creative but inevitably partial depiction of the world. Our beliefs about the natural world are not fully determined by the evidence available to us; they are said to be underdetermined. And once it is accepted that our scientific theories and categorizations are underdetermined by the evidence it becomes important to identify the other factors which incline us towards the particular beliefs we hold. In this way, sociological studies can take us into the heart of science. And if social factors are, therefore, a constitutive part of science it is no longer possible to treat science idealistically.

Results from a large number of case studies indicate how social factors – both small–scale and large–scale – are influential in deciding which beliefs come to be held. It has been proposed, for example, that some scientists' social class background predisposes them to view the natural world in a certain way (MacKenzie, 1978). Other studies (for example, Collins, 1975) have indicated that apparently technical matters like the reliability of experimental results are settled by negotiations between scientists. The outcome of these negotiations depends on scientists' assessments of each other's competence and reliability as well as on the technical matters at hand. The idea that scientific beliefs are always underdetermined by the available evidence, and the consequent acknowledgement that current scientific beliefs are not an unquestionable reflection of nature, offers an enormously wide scope for sociological analyses of science. Sociologists can inquire into the social construction of the very details of scientific knowledge claims. Such arguments are clearly antithetical to the notion that scientific knowledge unfolds in a pre–set, asocial manner. Even the status of the facts of science is believed to be constructed by the activity of scientists.

Yet while both these forms of social analysis of science are concerned with the connection between social processes and the development of science, the two sets of analysts have remained curiously insulated from each other. No doubt this is partly because they have focused their interests on different sites. On the one hand, constructionists have selected pure science as their research focus because this has traditionally been seen as the most disinterested and non–social form of science. It has accordingly been vital to discover social influences at the epistemological heart of science. Political economic analysts have studied areas like pharmaceuticals and information technology where the social consequences resulting from scientists' pursuit of particular lines of work have been most marked. They have wished to argue that, with different social priorities, other scientific knowledge could have been generated.

In addition to the insulation created by this division of labour, a further issue separates these schools. Since much of the critical force of the latter view derives from the fact that it regards scientific and technical knowledge as effectual and forceful (forceful enough to be, on occasions, socially harmful), its proponents see little to be gained by embracing the idea that scientific knowledge is just a *construct*. This source of difference makes it clear that the social constructionist position is even more radical in its opposition to idealistic interpretations of science than is political economy.

An Outline of the Text

The apparent conflict between these two interpretations is a central concern of this book. The chapters are organized to include material which corresponds with the preferred foci of both these approaches, ranging from basic science to the most obviously commercial international technologies as well as including such intermediate cases as applied science and state–funded research. The text has twin aims: to provide a sociological review of the involvement of science and technology in contemporary social change in the First and Third Worlds, and to assess the relative strengths of the social construction and political economy interpretations.

The first two chapters concentrate on the institution of science as we currently know it. Chapter 1 examines the nature of scientific knowledge. It considers in what sense scientific theories are underdetermined by the evidence; how defenders of scientific rationality have sought to respond to this indeterminacy; and how successful sociologists have been in systematically documenting the social influences at work in the social construction of scientific knowledge. This theme is developed in Chapter 2 where science is depicted as a social movement. Scientists' strategies for gaining authority for science and in establishing suitable organizational structures for the pursuit of science are analysed.

The next pair of chapters provides a discussion of the connections between science, technology, the state and the economy in industrialized nations, using the United Kingdom as the principal example. In Chapter 3 science is considered as a form of work. The nature of scientific labour in the commercial and the higher education sectors is reviewed. The system of state support for research is examined, and political aspects of this system are discussed. Chapter 4 is largely concerned with the benefits science and technology offer and convey to the economy. The nature of innovation in Western economies is studied as well as the role of state–supported research.

While the material covered in Chapters 1 and 2 is readily assimilated to a constructionist view, Chapters 3 and 4 concentrate on more openly political and economic issues. Chapter 5 offers the opportunity to compare these approaches explicitly. It deals with technical change and the sociology of technology and, to a large part, is dedicated to the analysis of one extremely pervasive technology: weaponry and military equipment. This chapter provides the opportunity to test the rival interpretations by asking whether social construction is important in one conspicuously political arena of technical decision–making.

In Chapter 6 these arguments about science and technology are applied to the case of the underdeveloped world. The consequences of technology transfer and technological dependency are considered. Case studies reveal the ways in which Western technologies are often inappropriate to underdeveloped economies. The viability of suggested counter–measures, such as the introduction of 'intermediate technology', is also reviewed. Chapter 7 draws together the themes of the book and reconsiders the relationship between the social construction and political economy approaches to the study of science. It is argued that a coherent sociological view of science and technology is required for an understanding of the way they figure in contemporary social change.

Note: Introduction

1   This tale was drawn to my attention in a lecture on Victorian science entitled 'Evolution and the popular press in early Victorian England' delivered by J. A. Secord at Queen's University, Belfast, 6 May 1987.

[1]

The Authority of Science: Knowledge, Truth and Reality

Introduction

In the introduction to this book the idea was put forward that scientific knowledge should be regarded as socially constructed. This idea will be explored in greater detail in this chapter; objections to this view will be considered as well. Some important implications of the social constructionist view will also be examined; particularly, it will be necessary to ask whether a socially constructed science can still count as valid knowledge and what consequences the social constructionist view holds for the authority claimed by scientists.

To begin with we need an account of what is meant by the social construction of knowledge. The most well known account of the construction of social reality is to be found in the work of Berger and Luckmann (1971). These authors began with an apparent paradox. From a disinterested standpoint it appears that many socially important beliefs differ vastly from one culture to another; that is, they appear merely to be conventions. Yet for members of those cultures, the beliefs they hold are enormously real and seem simply to reflect the way the world is. As a prime example Berger and Luckmann cite beliefs about sexual morality. In most cultures there is a strong sense of correct sexual conduct and behaviour in conformity with the approved morality is held to be natural. But, these authors assert, virtually every conceivable sexual practice is regarded as normal somewhere. Hence the feeling that one's own practices are *the* natural form of sexual conduct cannot be straightforwardly accepted.

This state of affairs holds for other beliefs too, beliefs about morality, religion, educational practices, gender roles and so on. The vast diversity of beliefs which cultural anthropologists have been able to record appears to conflict markedly with people's intuitions that only their society's (or their subculture's) practices are justifiable and satisfactory. Whilst actually being hugely variable from one society to another, beliefs generally seem completely beyond question to the members of any particular culture. To Berger and Luckmann this indicates that a prime task for sociology must be to understand how widely differing beliefs are generated and how these beliefs attain their compelling quality for the members of a society. Accordingly, these authors state (1971, p. 15) that:

> the sociology of knowledge must concern itself with whatever passes for 'knowledge' in a society, regardless of the ultimate validity or invalidity (by whatever criteria) of such 'knowledge'. And in so far as all human 'knowledge' is developed, transmitted and maintained in social situations, the sociology of knowledge must seek to understand the processes by which this is done in such a way that a taken–for–granted 'reality' congeals for the man in the street. In other words, we contend that *the sociology of knowledge is concerned with the analysis of the social construction of reality* [italics in original].

Berger and Luckmann do not directly address the issue of the status of scientific knowledge; instead they concentrate on beliefs which are of widespread importance in everyday life like morality and religion. But there is clearly a potential problem with the analysis of science hinted at in the above quotation since science is aimed at transcending 'whatever passes for "knowledge"' and at establishing findings of 'ultimate validity'. In the face of such an objective these authors' statements appear distinctly equivocal. Advocates of science would respond to Berger and Luckmann by claiming that it is an exceptional form of knowledge. Indeed, unless it is an exception, scientific knowledge would be in danger of losing the kind of authority which was described in the Introduction.

17

The Status of Scientific Knowledge

The issue of the status of scientific knowledge is best approached through a direct consideration of the foundations of science's authority. There are two fundamental appeals which can be made in defence of its authority. Science may be said to be an exceptional form of knowledge because it is based on facts or because it is uniquely true – its truth being indicated by a variety of forms of evidence including its utility. Other important areas of human belief could be said to lack these forms of validation since, for example, moral convictions are known to be evaluative and are not based simply on generalizations from observed instances of moral conduct. Since moral beliefs do not reflect reality but are intended to dictate to it, sociological analysts are comparatively at ease with the idea that opinions may vary from one culture to another. It is said, therefore, that such beliefs are culturally relative. Scientific ideas, on the other hand, are thought to be validated by their correspondence with the natural world. Under these circumstances there ought not to be room for differences of opinion; diverging interpretations would have to be put down to one person's error or to uncertainty.

Our confidence in the ability of facts to validate scientific beliefs often draws support from the analogy with perception. Facts are often taken to be evident in the same way that the things we see are evident; we are passive recipients of knowledge in the same way as we receive the evidence of our senses. However, this analogy ironically works to undermine the case which it is supposed to support since perception is far from the passive operation that this argument implies. Observation goes far beyond the simple bombardment of our retinas with light, since objects which we 'see' as the same present very different appearances to the eye. As sunlight is reduced by a passing cloud, the colour and appearance of objects – as measured, for example, by a light meter – change yet we regard the image of objects as staying the same. We can retain the constancy of an object despite approaching it from many different angles. Equally, of all the potentially available visual information available at any time we use only a minute amount. Seeing is usually the imposition of interpretative schemas on to the available information; observation is a fusion of interpretation and the reception of light. The extent to which humans customarily depend on interpretation in seeing is demonstrable through *trompes l'oeil* (trick paintings which convince us, for instance, that a solid brick façade has windows in it or that a drinks cabinet is really a set of bookshelves) or through experiments in the psychology of perception where, for example, we see lines drawn on a plane surface as a three–dimensional object (Collins, 1983a, pp. 88–90).

Just as perceptions have to be worked at, so do observations of facts. An unskilled observer allowed to wander in a forest will experience great difficulty in 'observing' the different kinds of trees, for there are likely to be as many variations between members of one species as there are between the differing types. The difficulties would be magnified if the observation were done over a protracted time period so that buds came and went, fruits appeared or leaves fell. Skilled sylviculturalists on the other hand would not only immediately see the kinds of tree but would be able to observe higher-level kinds of similarity between different species; see signs of health and vitality (or their absence); and see whether the season was late or early. This is not of course to throw any practical doubt on the usefulness of foresters' perceptions. It is, however, to cast suspicion on the idea that science is valid because scientists see the world *plainly*. All useful seeing is skilled seeing.

However, the status of scientific facts is even more complex than this analogy with perception reveals. Many scientific observations are made with machines; one observes an electric current or radiation from space not with one's senses but with an instrument. Scientific observations, particularly in the context of experiments, also have to be separated out from chance occurrences. Even scientific apparatus appears to be inhabited by gremlins, and the observation of a change in a meter reading or of blips on a pen–chart has to be divided into real facts and mere artefacts. Furthermore, the frontier of things which count as factual observations tends to shift as scientific ideas change, so that what would at one time have been regarded as hypothetical images from an experimental tele-scope, for example – comes later to be regarded as unproblematic observation. Finally, the facts of interest to scientists are commonly not isolated facts but facts about classes of things such as that dandelions have runcinate leaves. Yet no one could have observed all dandelions, particularly not prehistoric or future ones, and so there is something undeniably conjectural about such factual claims.

In some ways, however, it might be argued that these objections to the idea of the primacy of observation are contrived; they are not real difficulties but are made–up problems. Academics, it might be said, often appear to be attracted to made-up problems like the stereotypical philosopher's anxiety that all life might be a dream. But these problems are not fictional ones; they do have a relevance for at least two reasons. First, the kind of certainty which the factual foundation of science is supposed to offer is an 'in principle' kind. The factual basis of science should secure it even against imagined challenges or 'thought experiments'. Turning back for the moment to the implied contrast with moral relativism, we can easily imagine that claims to the effect that sexual practices or gender roles might be different would seem far–fetched in a morally closed society. Suggestions that women might participate in government might be seen as all too dreamlike. Yet we would not respect the argument that these challenges to morality were 'contrived' or 'fanciful'. Absolute moral certainty crumbles once one accepts the principle that all our beliefs could reasonably be otherwise. So too does the naïve confidence in the factual basis of scientific belief. Second, and more persuasively, it is anyway the case that just these sorts of issues do get questioned in scientific disputes. The study of controversies has been one of the principal areas of investigation in recent years by sociologists of science precisely because the apparent certainty of observation is undermined in these circumstances (Collins, 1981; Mulkay, 1980). In a recent study of solar neutrinos Pinch (1981) has provided a valuable example of these kind of factors. Solar neutrinos are curious, virtually massless and chargeless particles, generated as a by–product of the nuclear reactions in the sun, which travel through space to the earth. They are of interest because of the information they can impart about the nature of the sun. Pinch reports however that a conflict exists between the expected flux of these particles, based on other sources of knowledge about the form of reaction proceeding in the sun, and the actual measures of the flux. In such a clash of interpretations the observation of the flux at the earth's surface might be expected to take precedence; it is after all an observation.

But the nature of the observational activity makes this preference less obviously justifiable. Being massless and chargeless the particles cannot be observed in any routine sense; rather they can only be detected because the neutrinos are believed to cause a specific isotope of chlorine to transmute into an isotope of argon. This latter isotope is radioactive and the number of incoming neutrinos can be estimated on the basis of the amount of radioactive argon formed. The difficulty is that the target material, containing the chlorine, has to be of a vast size to generate a measurable number of transmutations. Around 100,000 gallons of the target material have to be contained in a tank. The tank itself must be housed underground in order to avoid the influence of other particles striking the earth which might affect the chlorine; these are believed

20

to be effectively filtered out by the burial beneath rock. Still, out of this vast mass of target only a few atoms will be affected, and so the argon has to be flushed out of the tank and concentrated before it can be accurately measured. The observation of the neutrinos is a far from straightforward process. For one thing the whole business of observing is dependent on pre–existing theories: theories about radiation counters for measuring the argon; theories about transmutation for reliably generating the argon from chlorine; and theories about the *neutrinos themselves* which account for their unique ability to penetrate the rock above the tank. There are huge areas for potential error, for example in the flushing of the argon out of the tank. As one interviewed scientist explained, 'The experiment is so complicated and it's so big ... there must be something in this experiment that's not coming out right' (Pinch, 1981, p. 140). Other participants made an even more explicit point about the dependence of the observation on other untestable assumptions (Pinch, 1981, p. 140): 'He has to make estimates of the cosmic–ray background, for example ... It has to be calculated, it's not actually measured...I think he has to extrapolate' (ellipsis in original). The opacity of the rock above the tank to other forms of radiation is queried by this scientist. All through, the observation is based on interpretations alongside the brute observing. Even the final counting of the number of argon atoms is done with a machine and not with one's senses.

Overall, therefore, this observation of facts about neutrinos is not free of any of the limitations on scientific perception listed above; in this instance those limitations are not contrived. The observation is based on machines and not on one's senses and it depends on prior theories (about, for example, other cosmic ray phenomena). Furthermore, the claim is about solar neutrinos *per se* rather than about the flux on any particular day; yet the expense of the experiment means that it can be done only in one particular location and for relatively short periods of time; so the practical problem resembles the case, mentioned earlier, of observing all dandelion leaves. Finally the success of the observation depends on the routine correctness of the steps in the experiment even when they are as problematic as looking for a few atoms in 100,000 gallons. As a nuclear physicist exclaimed (Pinch, 1981, p. 140): 'I find it very hard to see how we can pick up a few of these argon–37's, and maybe you are losing some of them...I just sit back here and think of this enormous big tank...and you are tying to flush out a few atoms' (ellipsis in original).

Such problems are not confined to sciences dealing with invisible entities either. In the course of a series of interviews conducted by the author with life scientists researching the origin of life on earth, a palaeontologist stated that biologists unaccustomed to the difficulties of observing small traces of life in very old rocks had mistakenly 'seen' signs of organic life.[1] He claimed (cited in Yearley, 1987a, p. 85):

> the people that did the work had never looked at the sort of, never looked down a microscope before, couldn't have done because there's paper tissues, there's bits off their jumper, hair and dandruff 'n' [laughing] all the pollen that you get out of the air in there.

No doubt every scientist has, in the most literal sense, looked down a microscope, but the speaker here discounts the kind of 'seeing' that other scientists have done. Seeing in science is evidently very difficult especially at the forefront of knowledge. Before the scientist knows what he or she is looking for it is very hard to spot it unequivocally. These biologists were looking for early life and found organic molecules in early rocks, but according to the palaeontologist above this was insufficient. Of course once the majority of scientists have agreed about the kind of life that is being looked for, it becomes relatively easy for the practised observer to identify it. This, however, is the crucial issue since it is precisely at the forefront where new knowledge is being established that observation should matter. Ironically, it is just at this point that observation encounters great practical difficulties. It supplies information but not indisputable facts. Just because science is based on observation, that does not mean that scientific knowledge is unquestionable or undeniably right. It could be observationally based and constructed at the same time. Current scientific beliefs may just be one way of construing the information we receive from the natural world.

The alternative argument for the non–constructed nature of scientific belief comes from the apparent truth of science. One could say that it is precisely because science works, for example that it allows us to develop new technologies, that we know it must be correct. Clearly the utility of science is a crucially important consideration, although, as will be seen in subsequent chapters, this utility is not as straightforward as might be thought. But if we return to the analogy with morals for a moment it will be easier to assess the strengths of this argument. While it would be correct to accept that the utility of science is good evidence for its cognitive worth, it is at best only *evidence* for the truth of scientific beliefs and not *proof*. In a directly analogous manner we find that throughout the world, as has been mentioned, a great variety of moral and religious belief systems are found to work. People behave in an orderly and emotionally satisfying way and that has always counted for them, if they have ever considered it worth thinking about, as evidence for the validity of their moral system. From an outside standpoint, however, we can say that its adequacy is no more than that – adequacy. It is not in any clear way evidence for the conclusive rightness of any particular system of morality. Equally, the utility and practical sufficiency of science can show only adequacy; we may believe that it hints at something more than this, at some kind of transcendental correctness, but this is a hunch. If one is going to take a disinterested look at the role of science in contemporary social development it is better not to stake anything on this hunch at the outset.

It can be seen, therefore, that this second anti–constructionist argument takes one of two major forms, neither of which should trouble the constructionist. The first form is that of an appeal to how certain we feel about science. Surely, it might be said, we would not feel this certain about a construct. But as we have seen, cultural anthropologists record that other peoples have equally strong feelings about moral or religious issues, and we do not regard those peoples' intuitions as indicating that their knowledge is really of a special order. Equally, we evidently used to feel strongly enough about religious heresy to burn people on account of it yet we now consider those intuitions to have been mistaken. Appeals to intuition seem to have been a bad guide in matters of epistemology. The second sort of claim – about evidence for the quality of science – actually comes very close to circularity. Commentators are often concerned to persuade us of the truth of science in order to get us to take science seriously – to take its advice or accept its authority. Science's truth is of interest because it gives us grounds for believing that the recommendations of science will work. Yet at the same time we are often invited to take the utility of science as 'proving' its truth. When these two halves of the argument are put together it appears that an abstract inquiry into the general truth of science is likely to be most unrewarding.

It should of course be borne in mind that to say that science is only adequate is not to belittle it. For one thing, to be adequate for the whole range of things it offers us is quite an achievement. Moreover, no other beliefs or theories could be any more than adequate in this sense either. The argument being advanced here is not that science is less true than other types of belief; it is rather that there is no point in rushing to give science more dignity than it currently holds by saying that, on top of all its other qualities, it is also true.

Revitalizing the Adequacy of Science

Up to this point the aim of the discussion has been to argue that the immediate and intuitive grounds for denying the constructed nature of scientific beliefs are not persuasive. Scientific beliefs can be about facts as well as being constructs; equally, they can be both adequate and constructed. But the argument has not rested there. Many commentators, mostly philosophers and scientists, have felt that science has to be more than a construct or at least a very special kind of construct. A number of authors have attempted to spell out formally the basis for this feeling that science is special. By and large they have adopted one or other of two strategies for identifying the special features of science. Either they have sought specific rules for the rational manipulation of constructs or they have looked for the particular values which inform the selection of constructs in science. Both these strategies centre on the ways in which scientists handle conflicting constructs; they focus on the resolution or closure of competing interpretations. Closure and resolution will turn out to be key themes. It is now worthwhile looking at these two strategies in turn.

Authors who adopt the first approach are often described as 'conventionalists'. This name indicates that they accept that scientific beliefs are in many respects a stylized and conventionalized portrayal of the natural world. For example, throughout the eighteenth century the influence of Newton's work was strongly felt in all the areas we would now regard as sciences. Accordingly the natural and experimental philosophers of the time generally looked for explanations of natural phenomena in terms of interactions between underlying particles. The reflection of a beam of light off a mirror could be accounted for in terms of light particles rebounding off the mirror's surface like a succession of balls off a wall. The interpretation of ill–understood natural phenomena as though they were made up of the action of particles was a convention. It went beyond pure observation; indeed it made sensible observation possible since it directed people towards appropriate objects and experiments to look at.

The next step taken by conventionalist writers is to argue that what is special about science is the attitude adopted towards these conventions. Scientists are critical of the conventions they subscribe to at the moment. Conventions are only provisionally accepted. And there are particular circumstances under which scientists are willing to move from one convention to another. There are, so to speak, definite rules for moving from one interpretative set to an alternative. It is these rules which lend science its uniqueness and which lift the conventions of science above the status of mere constructs.

This position is clearly quite different from the ideal of science as a growing stock of observational knowledge. Such a view of the simple, incremental growth of science had anyway been disrupted by the argument, famously associated with the historian Kuhn (1970) but noted also by other authors (Hanson, 1965; Toulmin, 1961), that scientific knowledge tends to grow in an uneven way. Major cognitive changes are followed by long periods of quiet growth which, in turn, are interrupted by another cognitive revolution. Thus, for centuries people successfully believed that the sun rotated around the earth and managed to generate complex and accurate accounts of the planets' motions, the occasions of eclipses and so on. This whole tradition of work was interrupted by a cognitive change in which the sun was moved to the centre and the earth was relegated to the status of a middling planet. Such immense alterations of opinion look rather like a change in social constructs; they might be thought to resemble a society's religious transformation (the Roman adoption of Christianity) or a change in morality (concerning women's right to choose a marriage partner). Conventionalists accept that these changes are indeed changes in convention, or research programme as they term scientists' conventions. But they argue that these changes are rationally compelling. They are not just changes but changes for the better. And the fact that they are for the better can be demonstrated with respect to rules for choosing between research programmes. An example will usefully illustrate what is at stake here.

At the close of the eighteenth century and the beginning of the nineteenth there was great interest in the natural history of the earth. It was widely accepted that a great many rocks occurred in strata and that these strata corresponded to the order of formation of the rock types. Other rocks, such as granites, were said to be unstratified. They did not appear to exhibit such a regular ordering and even seemed to cut across the other formations. There existed two predominant interpretations of the formation of rocks; some

25

investigators (often referred to as 'Neptunians') argued that the prime factor in shaping the development of the earth was water. The rocks had been successively deposited from a primeval ocean, and this lent them their regular order. Such authors ascribed the unstratified rocks to short–term, local inundations.

Opposing views were put forward by scientists (or natural philosophers as they would have styled themselves) who concentrated on unstratified rocks. According to these 'Plutonists', such rocks had resulted from the cooling down of molten material which had flowed into place. There were thus two research programmes. One set of people claimed that rocks were basically deposits from oceans and that water was the most important geological agent. Others insisted that the most general model was the cooling of molten rocks and that heat was the predominant influence. Several points can be displayed from this short account of this episode. Neither side's views were wholly the result of observation, although supporters of both sides were careful to make observations. By necessity, much geological information is buried beneath the ground, so that all the facts were available only by arguing by extension from what could be seen. Nobody had seen the formation of much rock in any case. On one side people had witnessed occasional volcanic eruptions and concluded that this was symptomatic of the formation of new rock from the molten state. Others had seen the build-up of sediments at river mouths and in lakes and thus seen the beginning of something that looked (somewhat) like mudstone or clay. But the formation of the earth's rocks had not been observed. In this sense both interpretations were conventions: they went beyond the facts at hand.

It should be stressed that both sides had good evidence; in this sense both cases were scientific. But, equally, both sides had weaknesses. There were limestones and sandstones which looked like hardened sediments. These argued for a watery origin for rocks. Yet rocks of this type were found high up on dry land, sometimes thousands of feet above sea level, and bearing all the marks of having been marine sediments. It was frequently asked where all the water had gone from the oceans. On the other hand, advocates of the importance of heat were faced with the problem that volcanic rocks formed today resembled only a small minority of the rocks on the earth's surface. Sometimes limestone rocks were found in close association with the rocks that were supposed to have formed from a melt, and yet limestones were known to decompose when heated. Worse still, in experimental trials in which common unstratifed rocks were melted down and then cooled, the molten rocks turned into glassy substances rather than solids with a stony appearance. Both sides offered wide explanatory power going beyond the facts at hand, both had strong supporting evidence but both also were confronted with anomalies.

The rationalist argument is that this is precisely the point at which the particular rationality of science is displayed. The resolution of this conflict of interpretations is the place where the specific characteristics of science are revealed. Each research programme has central tenets which are linked to actual observations through a series of supplementary claims. Scientists will tend to respond to the criticisms of their opponents by seeking to make sense of and explain the anomalies to which their opponents have drawn their attention. They will attempt to bring the anomalies into line with their central commitments. These philosophers accept that from a logical point of view some alteration in the peripheral band of supplementary claims can always be made to account for the anomaly; none the less as the use of these alibis builds up, one interpretation will tend to become more and more plausible at the expense of the other.

To illustrate what is meant here: the existence of limestone in close association with the unstratified rocks can be made sense of if we employ the additional idea that at the time the two rocks were some distance underground and that the pressure of burial prevented the limestone breaking down by giving off gas. An Edinburgh scientist, James Hall, went to the trouble of building a pressurized apparatus in order to show that at great pressure limestones can endure high heat without breaking down. Hall was able to add a new, supplementary proposition which preserved the theory of igneous formation in the face of an apparent anomaly (Yearley, 1984a, pp. 28–9). His experiment appeared to be widely accepted and thus provided a plausible defence of his central commitments. In this way, an apparent clash with known facts is tolerated and may spur scientists on to find new information which undermines the supposed counter-evidence. As scientists respond to each other's claims and counter-claims the relative merit or convincingness of the two sides changes. A second factor is also very significant: special importance is attached to the ability of one interpretation to anticipate some as yet unknown finding which the other does not predict. Many conventionalists claim that a great alteration in the plausibility of the competing sides is actually effected in this fashion. If, for example, the proponents of the influence of water had argued that the water had drained off into underground cavities and such cavities were found, then the prediction of a novel observation would have been especially deserving of favourable attention. Essentially, these philosophers argue that science is rational because the way in which choices are made between competing interpretations is bound by ground rules (like the injunction to value novel predictions over explanations of existing observations). As scientific work progresses, one interpretation will come to be favoured, and personnel will switch over to the research programme on the ascendant.

This modified rationalist account of science is both attractive and descriptively useful. It offers a view of the growth of science which does justice to the complexity of scientific argument. The question is whether this interpretation of the growth of scientific knowledge is sufficient to offer a prescription for the rational growth of science. This philosophers' account may be descriptively helpful but does it achieve the philosophical task it set out to perform? The rationalist argument is that rules are available for choosing between competing research programmes. There are two problems with the use of these rules. First, the rules face great practical difficulties. They appear to work best if there is an even contest between competing research programmes. Under these circumstances advocates of either side are free to respond to the new claims and counter-claims of their opponents. Suppose, to take the geological case cited above, the contestants had been free to develop their work and the advocates of rock formation by heat had achieved the following breakthroughs. They had succeeded in answering all the counter-claims of their opponents (for example, they had explained how rocks could form from a molten condition without forming a glass and they had shown how limestone and igneous rocks could occur in close proximity) *and* they had stated claims which their adversaries could not match. They might, for example, have been able to explain how volcanoes arose and how unstratified rocks like granites came into being whereas their opponents had to invoke *ad hoc* explanations to make sense of these phenomena. Under such circumstances the igneous theorists would be in a very strong position.

In practical terms, however, such a circumstance is unlikely. For one thing scientists have to decide for the present what is likely to be correct and what is not, and choose their work accordingly. They do not have the leisure to wait for the evidence to come out. Moreover, in any real scientific dispute the contest is almost certain not to be fair. It may simply be harder to get evidence for one theory than for another. Even two hundred years later we know very little about the earth's crust. Some sorts of evidence are just hard to come by. Career pressures or the availablity of jobs, material and funds may influence individual scientists as to where to work and with whom to co-operate. New scientists will emerge from a prolonged period of training during which they are likely to have been persuaded that the view of one school or the other is more plausible. In all these ways social and psychological pressures militate against the occurrence of a truly fair contest. The rationalists' own position is unstable in this regard since they are obliged to talk of scientists being persuaded to favour the ascending programme. We cannot tell for anyone who is persuaded whether it was good scientific evidence alone or that and practical considerations which decided her or him.

Rationalists may retort by claiming that their theory is in any case an 'in principle' idea and not subject to criticisms for impracticality. Impractical rules are not of much use of course, but even if one accepts the restriction to in principle matters the outlook is not good. As has been stated, in most actual scientific disputes we cannot wait for the contest to run its course. One interpretation will not decisively oust the other; rather there will be a balance of pros and cons. This situation of balance cannot be handled by these rules. But even in the imaginary case outlined above, the issue is not straightforward. The philosophy is supposed to offer a procedure for telling when one programme is rationally superior to the other; but even if all the odds favoured the igneous theorists, a staunch opponent could still insist that time would favour the competitor. Unless some arbitrary time limit is imposed the rationalist argument does not instruct us when people *must* switch over. Thus the very things which lend this theory its descriptive attractions – the fact that it talks of persuasion and plausibility – are its undoing, for it makes it too subject to psychological and social pressures influencing judgement and unable to offer a logical and rationally compelling point at which beliefs must be exchanged. The description is attractive (although not necesarily more so than one which generalized from sociological or historical case studies) but it is philosophically insufficient to recover absolute rationality in science.

Scientific Values

If rules do not accomplish the task set out by rationalist authors of setting science apart from other forms of belief, another basis will have to sought on which scientific authority can be erected. The other popular recourse for philosophical analysts has been values. Kuhn, whose early work was mentioned above, sought to reduce or overcome the relativistic consequences of these earlier studies by suggesting that scientists used a number of values for assessing the merits of rival scientific theories or research programmes. He proposed (1977, p. 322) that scientists prize highly the following five 'standard criteria for evaluating the adequacy of a theory': accuracy, consistency, scope, simplicity and fruitfulness. In assessing theories scientists will, on this view (1977, pp. 321–2), evaluate them along the following dimensions:

(1) The 'consequences deducible from a theory should be in demonstrated agreement with the results of existing experiments and observations'.

(2) The theory ought to be consistent internally and 'also with other currently accepted theories applicable to related aspects of nature'.

(3) The 'theory's consequences should extend far beyond the particular observations, laws, or subtheories it was initially designed to explain'.

(4) It ought to bring 'order to phenomena that in its absence would be individually isolated and, as a set, confused'.

(5) The 'theory should be fruitful of new research findings'.

Kuhn argues that scientists recognize that these features are desirable in scientific knowledge. To put it in the rhetoric of cereal-packet competitions, they use their 'skill and judgement' to assess the relative merits of contending interpretations in the light of these values. The scientific community is the sole authority on the comparative standing of scientific ideas; the values which guide the growth of science are those which scientists collectively decide on. There is no other authority to which appeal can be made. As Kuhn states later in the same paragraph (1977, p. 322) 'they provide *the* shared basis for theory choice' (italics added). These criteria are just a distillation of what scientists are found to do. One could, in an equivalent way, set out criteria encapsulating the activities of post-expressionist painters or fashion designers.

In Kuhn's statements, however, there remains an uncertainty about the precise nature, source and status of these criteria. For one thing he accepts that the above is not an exhaustive listing; of these five values he says (1977, p. 321): 'I select five, not because they are exhaustive, but because they are individually important and collectively sufficiently varied to indicate what is at stake.' However, unless all the values could be listed it is hard to understand in what sense the values can be said to direct scientific decisions. Second, the status of the individual values is unclear. There is a tension between the view that they are simply generalizations about the values which scientists happen to honour – just as one might record the values recognized by Wall Street brokers – and the suggestion that they have some intrinsic logic or that they derive from some transcendental standard.

Partly in response to these ambiguities Newton-Smith (1981) has sought to provide a fuller defence of the use of values to preserve the rationality of science. His initial approach to the question differs from Kuhn's in that he begins with a realist interpretation of science. Newton-Smith is cautious in his realism. He claims that science is distinguished from most other forms of knowledge because it tends to get truer as it goes along. Still, we cannot accept that our beliefs at any particular time are the truth. Rather, we must accept the 'pessimistic induction' (1981, p. 14) that we will sooner or later abandon our current beliefs as untrue for, judging by the history of science, everything which we now believe is likely to turn out to be false in some regard. We can, though, pick out criteria which have been used in assessing scientific ideas and which we have good reasons for thinking are linked to an increase in truthfulness or, as he terms it, verisimilitude. Several of the criteria suggested by Newton-Smith are similar to those proposed by Kuhn. He lists a series of eight 'good-making features' of theories (1981, pp. 226–32). These are:

(1) That a theory should 'preserve the observational successes of its predecessors'.

(2) That a theory should be fertile in producing ideas for further inquiry.

(3) That a theory should have a good track record to date.

(4) That a theory should mesh with and support present neighbouring theories.

(5) That theories should be 'smooth', meaning that it should be possible easily to adjust the theory in the light of anomalies which are bound to emerge.

(6) That a theory should be internally consistent.

(7) That theories should be compatible 'with well-grounded metaphysical beliefs'; that is, theories should accord with the same metaphysical assumptions as sustain the rest of science.

(8) Although hesitant because of the ambiguity of this criterion, it is probably beneficial for theories to be simple.

What is significant about this list of criteria is not just the individual recommendations but the claim that the values each have a double justification. Newton-Smith asserts that they are both the criteria by which scientists *do* judge and criteria which can be shown to be *rational* in the light of the assumed goal of science, that is to become truer and truer. Thus, theories should be compatible with widely adopted metaphysical assumptions because it is very hard to see how science could be becoming more correct if major sections of it depended on conflicting metaphysics. If a new physical theory, for example, meant that while biology required the universe to be one way, physics entailed another ordering, that would be a retrograde movement.

Newton-Smith thus seeks to tackle Kuhn's problem head on; his theory is avowedly empirical and normative. It is an account of the values which scientists as a matter of fact generally do take into account *and* it is a demonstration of why scientists are right to honour those values. It is this latter aspect which would potentially allow Newton-Smith to claim that science is rational and that scientific knowledge is uniquely authoritative. But how satisfactory is this normative element? As Newton-Smith himself makes clear, none of the criteria is inviolable. On occasions some values may have to be subordinated to others. For example, a theory (T_1) with a poor track record may be preferable to some other theory (T_2) because of (T_1)'s assessment on the other values even though (T_2) has a better track record. In the great majority of scientific decisions therefore a judgement will have to be made about the merits of different theories' 'scores' on the eight values. And with these eight criteria to be taken into account the scores can be totted up in very many different ways. Just using the criteria will thus demand a huge exercise of judgement by scientists. But the situation is even more complex than this, for the criteria are not automatic in their application. Taking criterion (4), for instance, meshing with and supporting neighbouring theories is a far from simple requirement. Which are the neighbouring theories? Looking back to the geological dispute outlined earlier, it would be evident for supporters of the Plutonist position that the study of volcanic activity was a field neighbouring the study of the formation of rocks. For Neptunians, however, the study of volcanoes would be only very remotely connected to the issue of the origin of rocks. But even if neighbours could be uncontentiously identified it would still be unclear how to evaluate the degree of support given to those neighbouring theories. Is it better to lend a great deal of support to a few neighbouring theories or to lend some support to a lot of neighbours? When viewed in this way it appears that Newton-Smith's approach is subject to the same practical limitations as those Kuhn (1977, p. 324) admitted for his own, since

> When scientists must choose between competing theories, two men fully committed to the same list of criteria for choice may nevertheless reach different conclusions... With respect to divergences of this sort, no set of choice criteria yet proposed is of any use. One can explain, as the historian characteristically does, why particular men made particular choices at particular times. But for that purpose one must go beyond the list of shared criteria to characteristics of the individuals who make the choice.

Newton-Smith took Kuhn to task (1981, pp. 122–4) for not rooting his proposed values in the rational requirements of science. For him, Kuhn was making too weak a case for science by implying that the five Kuhnian values were just a statement of how scientists conducted themselves. As a realist, Newton-Smith cannot allow that science comprises a set of values or criteria one can choose to follow or not. The values must not just be a convention; they have to be the real values for getting on best in describing the world. But, as we have seen, the practical normative force of the proposed eight values is much less than Newton-Smith would seem to require.

One may accept that there is a certain plausibility to the values. They may describe the kinds of considerations which scientists appear to have in mind when selecting theories; they may even strike us as the kind of consideration which scientists ought to have in mind. But, unless we have good reasons for thinking that the values direct scientific choices in a strong sense, this normative force is of limited consequence. Providing a list of values which scientists should honour but which, in practice, does not constrain scientific choice at all closely, does rather little to revitalize the authority of any specific scientific judgements. Newton-Smith supplies us with general grounds for thinking that science as a whole is a reasonable undertaking but he does not reassure us that any particular scientific judgement could not reasonably have come out differently. By listing his suggested criteria in a chapter entitled 'Scientific method', Newton-Smith might be seen as implying that the criteria can be used as something like a recipe for scientific rationality. It should now be clear that they cannot serve in this capacity.

Sociological Studies of the Construction of Scientific Knowledge

At the very beginning of this chapter and subsequently, in discussing the example of how neutrinos are observed, reference was made to sociological studies of the construction of knowledge. In the last fifteen years a large number of such studies have been undertaken (see Collins, 1983b; Mulkay, 1980; Shapin, 1982), the majority of which have concentrated on scientific debates and controversies. Under such circumstances there are generally two groups of scientists who are competing to have their interpretations of nature accepted. Since these groups are composed of recognized scientists one cannot argue that their differing interpretations could both be due to the same evidence from the natural world alone. The very existence of disagreement shows that beliefs are underdetermined by the evidence available. Just like the philosophers described in the last two sections, sociologists have been interested to determine how such disagreements are resolved. But instead of looking for a set of rules or a clutch of values which bring about this resolution, sociologists have sought, in the words of Kuhn cited above, to explain 'why particular men made particular choices at particular times'. That is, rather than look for a scheme which makes the choice appear rational, sociologists have studied scientific choices in the same way as other choices (about voting, divorce, career moves and so on) are studied. Such an approach is often said to be naturalistic. And because scientists are social beings – they commonly work socially in laboratories; they attend social functions like conferences in the course of their work; and they make their careers in the social structure of the scientific profession – their decisions are made in a social context also.

Such an approach to the study of scientific knowledge has led to some misgivings. Some supporters of the authority of science regard this naturalism as liable to compromise science. It is suggested that most of the time scientists make their decisions on rational grounds. They follow internal criteria for choosing between theories. Occasionally, however, scientists depart from such internal considerations and place external factors first: scientists may, for example, prefer a theory which reinforces their political or religious views over a better theory which is incompatible with those views. It appears, for example, that the mid-nineteenth-century British geologist Lyell insisted that humankind was the result of a separate creation even though he accepted the evolutionary descent of all other life. One might conclude that his religious preconceptions overrode his commitment to internal, scientific grounds for evaluating theories of mammalian descent (Bowler, 1984, p. 221).[2] Those who argue in this way equate occasions when sociological explanation is called for with occasions when science is subverted. Social factors may be called on to explain error since scientists are subject to human frailties; it is for this reason that the rationalist philosopher Lakatos (1978, p. 178) states that 'Because of the imperfection of the scientists, some of the actual history is a caricature of its rational reconstruction.' On this view, sociology deals with the occasional instances of 'imperfection'.

Ironically sociologists have sometimes lent weight to this interpretation. One study cited very frequently is an early essay on the socioeconomic roots of Newton's physics by a Soviet analyst, Hessen (1931). Stated with all possible brevity, Hessen's argument was that the agenda and contents of Newton's work were determined by his socioeconomic environment. Early capitalist development required certain forms of knowledge and Newton supplied them. This argument became very influential because of the challenge it provided to customary ways of thinking about the history of science (Graham, 1985). But, in other respects, it is an unfortunate representative for sociology to have acquired. At an abstract level Hessen's claim is that external factors (in the sense outlined above) rather than internal ones explain the contents of Newton's knowledge. Curiously, therefore, Hessen becomes an ally of Lakatos in so far as he equates sociological explanation with external factors. That this equation of 'sociological' with 'external' persists is attested by the following remark of Newton-Smith's (1981, p. 266):

the rationalist regards the history of science as constituting, by and large, progress towards the goal [of better knowledge]. The main explanatory role is accorded to internal factors. External factors such as the social conditions of the times or the psychology of the individuals involved come in only when there is a deviation from the norms implicit in the rational model.

Recent sociologists of science have not, however, wished to argue that internal factors are overridden by gross social factors. Science is very often a highly insulated, arcane pursuit. The study of neutrinos, for example, is hardly likely to excite many religious or political bodies, at least not in the same way as theories about human origins. People may become angry that so much money is spent on building a neutrino detector, but their annoyance will not be related to the 'internal' issues. They are as likely to be incensed by the unsuccessful pursuit of neutrinos as by a successful hunt. Thus the idea that broad social factors (like religious persuasions) directly determine the content of scientific knowledge is, with the exception of a small number of areas, becoming less and less plausible. Increasingly, scientific knowledge is constructed by small numbers of specialized workers. They are often very effectively insulated from broad social forces. But their construction of knowledge still goes on socially. To invoke Kuhn again, one can still examine the construction of knowledge by particular people in a particular social context. Or to put it in Newton-Smith's terms, one can study the use of 'internal' factors from a sociological point of view.

The 'Internal' Sociology of Scientific Controversy

By considering a case study concerning a dispute in physics we can see that a fully sociological analysis of scientific controversy is not limited to gross external factors. The dispute studied by Collins (1975) centred on an attempt to detect gravity waves. According to Einstein's general theory this type of radiation should sometimes be emitted by massive objects: for example, if a star were to collapse. However, the amount of radiation arriving at the earth would be very small and detection correspondingly difficult. In the late 1950s one physicist set about trying to build a detector which would be sufficiently sensitive to pick up gravity waves arriving at the earth. His apparatus had at the same time to be massively large and extremely sensitive. Only a huge piece of metal would be massive enough to be affected by the flux of radiation which reaches the earth and then it would move only by a microscopic amount. The detector bar thus had to be mounted in a vacuum and insulated from all other likely forces which might cause slight movement such as traffic passing the laboratory. Finally, the microscopic movements had to be observed, not of course by eye but electronically.

After more than a decade's work the pioneering physicist claimed to have successfully detected the radiation. Other scientists then decided to build their own detectors to replicate his finding. Some of these other detectors did not record any radiation, and a scientific disagreement ensued. Both sets of scientists were keen to do observations, but the observations they made did not decisively settle the dispute. Those, like the original inquirer, who recorded gravity waves argued that their opponents' machines were too

36

insensitive. The opponents, for their part, argued that the apparent detection of radiation was due to artefacts of the apparatus. Electronic signals which supposedly registered the effect of gravitational radiation were, in their view, due to ordinary vibrations penetrating the equipment or to random variations in the electronic pick-ups. Both sides could quite reasonably defend their own position and explain away the claims of their adversaries.

In Collins's account it appears that the scientists are concerned about internal factors; they want to build a device which correctly measures gravitational radiation. But the technical evaluations they need to make are not separable from more obviously social judgements. Collins asked representatives from each of the experimenting groups for their estimation of the other groups' work. The remarks he recorded about one such group (referred to as W) ran as follows (1975, p. 212):

(a) ... that's why the W thing though it's very complicated has certain attributes so that if they see something, it's a little more believable ...

(b) They hope to get very high sensitivity but I don't believe them frankly. There are more subtle ways round it than brute force ...

(c) I think that the group at ... W ... are just out of their minds [original ellipsis].

Both the credibility of the group and that of the information it supplied were assessed in a variety of ways. Estimations of technical, 'internal' matters are not divisible from assessments of such 'social' attributes as trustworthiness, skill and practical good sense. Yet the decision whether to accept the claims of group W, for instance, is made by individual scientists. The great majority of scientists no doubt decide very carefully, conscientiously and reasonably. But their decisions are none the less practical decisions about what to believe, whose experiment to trust and whose word to accept.

There are two further aspects of scientists' decision–making which are characteristic of science. The first issue arises because scientists will tend to try to improve upon earlier apparatus. This means that, very often, the version of an experiment on which a repeat test is made differs from the original equipment. This may be because there is more credit to be got from devising a new apparatus than from merely copying or because testing on a different machine provides an element of cross-checking.

Either way, the practical upshot of proceeding in this way was a considerable variety of gravity wave detectors. Disagreements about the phenomenon of gravity waves were then attributed to variations in the detectors. One consequence of the pursuit of a 'better' machine is therefore that another level of judgement has to enter scientists' decision-making; they now have to assess the rival merits of differing sets of apparatus.

The complexity of scientific choice is compounded by a second sociological factor. Any specific research area is likely to be fronted by a limited number of scientists. In the case of gravity waves Collins reports thirteen centres in the whole world; for a technically difficult and enormously expensive project like the detection of neutrinos there are likely to be even fewer. And even in comparatively inexpensive areas of research there will be limited numbers of scientists working. As we have just seen, the small numbers do not mean that the scientists agree or think well of each other. They may not even be well acquainted, being separated by many thousands of miles. But the consequence is that rather few scientists make up their minds about any particular scientific fact or theory on behalf, as it were, of the scientific community. These groups have been termed 'core sets' (Collins, 1985a, p. 143).

The significance of these last two sociological observations is that they provide further reason for thinking that internal scientific assessments are inseparable from social evaluations. Scientists must choose not only between each other's results but between each other's apparatus. And the numbers of people involved are sufficiently small that they could not possibly do every test and counter-test nor check all their opponents' results before they begin to make their decisions. They need to choose what to believe and whom to credit before they can get around to deciding what the facts are. If the group at 'W' are 'just out of their minds', apparent facts arising from W are likely to be treated with great caution. Thus, in practice, there are no purely internal factors in the sense that Newton-Smith intends. Naturally enough, scientists attend to matters internal to their laboratory life, but those matters are a fusion of social, personal and technical considerations.

The Social Closure of Scientific Debate

In the last section we saw how scientific knowledge can be regarded as a social construction even if scientists involved in a dispute appear to heed only internal considerations. A scientific dispute has no logical end point. Any evidence for gravity waves could always be explained away or doubted by determined opponents. All the same, disputes do get settled. For the social constructionist, the settling of a dispute must be a social agreement. A dispute is settled when one interpretation triumphs. We may talk about this in much the same way as rationalist philosophers do. We might say that, for example, pro-wavers became convinced that their machines were unreliable and they came to believe that no gravitational radiation could be detected at the earth's surface. A conclusion to the dispute had been negotiated. The evidence plays a part in these negotiations but so do all manner of other considerations. The sociologist does not aim to dignify this coming to agreement by seeking a way to exhibit its rationality. Instead all the factors involved in the persuasion are treated equally.

Treating science in this way brings a further advantage since it allows the reopening of the question about the role of external factors in the construction of scientific knowledge. Once it is clear that internal factors are not asocial the formerly rigid distinction between internal and external factors crumbles. Factors which Newton-Smith would regard as external may influence science not only by overriding internal considerations, as was perhaps the case with Lyell, but also by playing an internal role. An example which illustrates this possibility comes from the history of brain science.

In early nineteenth-century Edinburgh there was great topical and scientific interest in the new science of phrenology. Phrenologists claimed that physically different parts of the brain corresponded to distinct attributes of personality and character. Accordingly 'diagnoses' of character could be read off from the physical shape of the brain. Areas of the brain were held to correspond to specific attributes such as a person's capacity for hope or his or her sexual desire. All things being equal, someone with large reserves of hope would have a large 'hope' organ. The brain's surface would therefore be expected to be bumpy, with the strongest character traits represented by the bumps. Phrenologists believed that the contours of the brain were reflected in those of the skull. Accordingly, psychological readings could be taken by carefully feeling a person's cranial contours.

This psychological procedure attracted many supporters. Great success was claimed by skilled phrenologists in reading the characters of people formerly unknown to them. Today phrenology is often held up for ridicule, but it must be realized that at the time it was a very reasonable scientific approach to the mind. Phrenology could, indeed, be held to be rather modern because it was avowedly materialistic; it identified the mind with the brain and brought the mind into the scientific field of inquiry. As a scientific theory it could be tested and refined.

In his celebrated study of the Edinburgh phrenology disputes, Shapin (1979) shows that a social and political campaign became associated with phrenological science. Phrenology was adopted by a group of reformers as the scientific basis for a programme of social improvement. Study of the brain would allow people's inherent qualities to be revealed; social change could then be planned to realize the potentialities of all classes of society. At the same time phrenology was opposed by various groups in what may be termed the Edinburgh 'establishment'. Some objected to its materialism, some to its potential to disrupt distinctions of class and some to its challenge to moral theories of the source of human action (Shapin, 1979, pp. 144–6).

The dispute was partly fought out in an overtly political way. Political, moral and religious debates between the sociopolitical groups ensued. But another strategy was also adopted. Attempts were made to test the validity of the phrenological claims, since, if they could be thrown into doubt, there would be no natural basis for the reform programme. One obvious route for this challenge to take was to penetrate inside the skull to see whether there was any basis in the appearance of the brain for the phrenologists' belief in distinct character organs. Shapin discusses four aspects of the brain which were closely observed in order to try to settle the dispute. The debate can be illustrated by concentrating on just one of these, the study of cerebral fibres.

According to phrenological observers white brain tissue was composed of fibres. In their drawings of brain structures the fibrous appearance is very clear. For them the fibres were straightforwardly visible. Anti-phrenologists in Edinburgh were by no means as sure. One of their number, John Gordon, claimed (Shapin, 1979, p. 159) that 'When we make a section of [white tissue] in any direction with a sharp scalpel, the surface of the section is perfectly smooth, and of a uniform colour. There is no appearance of any cells, or globules, or fibres whatever.' For Gordon the fibrous appearance was an artefact generated by the clumsy way in which phrenologists scraped the white tissue from the brain. Phrenologists, in reply, maintained that (Shapin, 1979, p.160) 'We seldom cut, but mostly scrape; because the substance, on account of its delicacy, when cut, does not show its Structure.' Both sets of scientists were intent on observing. Yet they persisted in seeing different things. Those differences were embedded in their naturalistic drawings. They saw the world, or at least the brain, differently.

Shapin's claim is that their different perceptions were structured by the different interests they had as observers. The phrenologists saw fibres as the real appearance because fibres were easily compatible with the idea that discrete brain parts served different functions. Their opponents saw the fibres as artefacts because it was felt that the white tissue was really undifferentiated just in the same way as the brain was largely undifferentiated. For Shapin this offers strong support for the proposal that an external factor (these scientists' commitment to the socio-political debate over the brain) had an impact on the very things they saw in the brain. The external factor was mediated through their evaluation of the phrenological theory of brain function. Political factors impinge on scientific observation not by overriding observers' commitment to science but by influencing their scientific practice. These scientists did not take positions on the phrenological controversy *despite* what they saw; what they saw was structured by their place in the controversy.

Shapin uses this case study as evidence for a general sociological theory of knowledge construction. According to this theory, knowledge is always generated by people in the context of their pursuit of some interest (Barnes, 1977, pp. 1–19; Yearley, 1984b, pp. 68–72). On some occasions the content of that knowledge may relate directly to their interest. The Edinburgh phrenologists, for example, produced knowledge which was very closely linked to their political interests. For them phrenology was a powerful legitimation for their political campaign. At other times the interest may sponsor the knowledge only indirectly; Shapin (1979, p. 146) reports that some established medics were opposed to phrenology not because they objected to the implications of the theory of brain function but because it threatened to hand over diagnostic skill to amateurs.

In many respects this theory goes further than the work of other social constructionists like Collins, since the contest between the interests not only accounts for the conflicting scientific beliefs; it should also allow the sociologist to explain the outcome of a dispute in terms of the triumphant interest. We would expect the downfall of phrenology as a body of scientific beliefs to occur when its social base was removed. However, there are two practical difficulties confronting this sociological theory. The first is a matter of evidence. No matter how many case studies have shown that beliefs are founded on social interests and that the belief which wins acceptance happens to be the one supported by the strongest interest, this would not prove that disinterested assessment was impossible. Just because the brain scientists disagreed about the cerebral fibres at that time, it is not clear that they could not have come to agree eventually. Given the practical constraints under which scientists work, social interests may be extremely influential in shaping their perceptions. But it is not certain that scientists are ultimately constrained by their interests. The second difficulty lies with the concept of interest itself. Any scientist is likely to have many interests: political interests, career interests, personal interests and so on. And even within the range of socio-political interests which interest theorists often stress (as Shapin does in the case of phrenology) there will be short-term and long-term considerations. Any actor can therefore be expected to have a number of interests at any particular time. Some will play a part in structuring his or her scientific beliefs; others may not. Once a variety of interests is admitted, the explanatory power of this general sociological theory diminishes, although the importance of particular case studies remains.

Conclusion

This chapter began by examining the claim that scientific knowledge is exceptional because it is not socially constructed. But initial arguments defending its supposed exceptional quality were subject to many weaknesses. Science is about natural reality but it still depends on human judgement and human conventions. Two attempts by philosophers to demonstrate that science was a special sort of convention were then considered. These were not satisfactory either because, although they showed ways in which scientific beliefs were reasonable, they did not show how the 'correct' choice between competing scientific beliefs could be made. Both the rules and the values which were supposed to direct scientific choice were indeterminate. None the less those philosophical analysts provided a useful descriptive vocabulary for discussing theory choice in science.

The failure to make out a case for the exceptional status of science left scientific knowledge open to sociological study. Through the case studies of gravitational radiation and of phrenology it has been shown how scientific knowledge is negotiated and constructed both within the relatively insulated social environment of the laboratory and in the context of political debate. The sociological study of the making of scientific knowledge breaks down the apparent divide between the internal and external factors affecting scientific knowledge. But the fact that science is socially constructed does not mean that it is indistinguishable from other facets of culture. The values proposed by Kuhn and by Newton-Smith give a useful sense of what science is like; by and large it is observational; by and large scientists pursue accuracy and so on. Just because science is socially constructed that does not mean that it cannot generally be told apart from art appreciation or pigeon fancying. The values describe science; they do not govern it. If we wish to understand why we have the scientific beliefs we currently hold, we cannot simply appeal to the values. We need to study the particular choices made by particular people. We must study the competition for credibility between scientists themselves and between the theories they support.

Finally, in regard to scientific authority, it is clear that although scientific knowledge is socially constructed it is the most authoritative account of the natural world we possess. Scientists are quite reasonably regarded as authorities on the natural world. But their authority is not absolute since it does not stem from indubitable knowledge. Scientific authority does not warrant the exaggerated respect associated (as was mentioned in the Introduction) with scientism. An appreciation of the sense in which scientific knowledge is socially constructed allows scientific authority to be regarded in the correct light. It is much closer to the authority of a skilled craftsman or legal adviser than the scientistic image implies. In the next chapter we will see how the social authority of science has been built up and maintained.

Notes: Chapter 1

1   I should like to express my thanks to the scientists who gave time to be interviewed in the course of this research and to the Economic and Social Research Council, which supported the project (grant A33250031).

2   It should be noted that the case of Lyell is being used here only in an illustrative sense. It could readily be argued that, at the time, there were reasonable grounds for believing that the human animal was radically distinct from other mammals and was therefore quite probably the product of a separate creation event.

[2]

Science as a Social Movement

Introduction: the General and Particular Authority of Science

The last chapter was concerned with the status of particular claims to scientific knowledge: for example, of assertions about the properties of solar neutrinos or about cranial contours. It was argued that although scientific beliefs are based on observation and experiment, the role played by scientists' judgements and skills cannot be eliminated. For this reason scientific knowledge cannot be insulated from its social context. Philosophers of science sought to limit this potential for social influence by providing rules or values which were supposed to direct scientific judgement. But this position was beset with problems, since neither the rules nor the values constrained scientific belief sufficiently closely to make it clear that any particular scientific belief is rationally compelling and fully authoritative. Scientists may legitimately disagree about the weight to be attached to the different values they use in assessing scientific knowledge. Thus, a sociological approach to understanding the development of particular scientific beliefs was advocated.

The sociologist's interest in scientific knowledge focuses on the way in which interpretations of scientific evidence are settled. In the last chapter this process of settlement was referred to as closure. But to regard the closure of scientific beliefs as a topic available to sociological study does not imply that we can refer to just any beliefs as scientific. Generally speaking, scientific beliefs are aimed at offering an explanatory understanding of the world. And in this sense the analysis of scientific belief in terms of scientists' values

44

has some virtue. Beliefs which are candidates for inclusion as parts of science will ordinarily exhibit the characteristics mentioned by Kuhn and Newton–Smith.

Thus, when the sociologist looks at particular scientific beliefs he or she often expects to see scientists competing to have their judgements and their versions of nature accepted. However, when the sociologist turns to science in general a rather different view is anticipated. Here we see a collection of beliefs with many similarities. Such beliefs can reasonably be characterized as empirically based, as aimed at accuracy and so on. At the same time as we acknowledge that particular scientific beliefs are conventions, socially shaped and provisional, we may regard science in general as the only authoritative, naturalistic way of interpreting the world. As was stated in the Introduction, science has largely displaced earlier exemplars as a generally authoritative way of interpreting the world. But how has this position of authority been attained and developed?

The aim of this chapter is to show that this question about the authority of science (rather than of particular scientific claims) can be answered in a similar way to the question of the standing of particular scientific knowledge claims. The growth in the authority of science can be understood as the outcome of competition to earn recognition for science as a form of knowledge. The establishment of science's authority can also be analysed as a process of closure, a closure of competition between forms of cognitive authority. A central analogy which will assist in making sense of this claim about the nature of scientific authority is that of science as a social movement. In the sociological literature the term 'social movement' is applied to informal organizations set up to pursue a goal or set of objectives. Current examples would include groups pursuing a change in the law relating to censorship or abortion or seeking a reform of religious practice. In many respects science can be viewed as an informal organization with cognitive goals. There are, of course, some weaknesses in this analogy, but this image will prove helpful in asking questions about the authority of science: questions concerning the way in which that authority was built up and maintained and about the organizational arrangements adopted by scientists.

The Professional Ideology of Science

In the course of their development social movements and other organizations commonly produce a favourable image of their objectives and a justificatory account of their activities. As was noted in the Introduction, there exists a widespread image of science as unquestionably correct knowledge. On this 'scientistic' interpretation, scientific knowledge is peculiarly authoritative. Science is regarded in this way because it is said to be based on observation and because it is impartial. Clearly, in general terms it is correct to say that science is based on the observation of the natural world, but even realists like Newton–Smith would no longer argue that it is absolutely observational. This scientistic view can therefore be regarded as a favoured presentation or image of science. Such images are often described as ideologies. It will prove useful to inquire into the use and development of the ideology of science.

In discussions of the broadcast image or ideology of social groups, the ideology is usually regarded as serving one of two purposes. It is seen either as promoting the interests of the group amongst outsiders or as heightening internal solidarity and helping to remove the internal strains from the enterprise (Gieryn, 1983, p. 782). In a recent study of a group of biochemists, Gilbert and Mulkay identified an interpretative device which appears to fulfil the latter function. These sociologists interviewed a number of scientists who were actively engaged in a controversy over the way in which biological cells generate energy. Representatives of both contending opinions spoke as though their own views arose directly from close observation of nature and careful experimentation. These scientists were, however, confronted by eminent scientists who disagreed with them and who, in turn, defended their beliefs by appeal to the same epistemological virtues. When asked about this divergence of views or when the topic of the conflicting beliefs came up in discussion the scientists commonly handled the interpretative problem in something like the way exhibited in the following statement taken from one of Gilbert and Mulkay's interviews (1984, pp. 92–3):

[scientific judgement depends on] intuition based on experience and, secondly, on one's feelings for the honesty and capacity of one's colleagues. I think that's a very important thing, which is very often unsaid in science, but I'm sure plays a very important determining role in the progress of ideas. I know it's so in scientific meetings. People will pay attention to some people and not to others. And sometimes it's a very false sort of thing, because it also has mixed up in it the whole thing of charisma and how nice a person is rather than how competent they are. So it's a somewhat unreliable guide, but I'm sure it plays an important part in determining the course of events. (*pause*) I think *ultimately* that science is so structured that none of those things are important and that what is important is scientific facts themselves, what comes out at the end [italics in original].

In this reply the biochemist details many aspects of scientific judgement which appear to depend on variable and unsystematic aspects of the scientific profession. He appears to give every support to a view of science in which personalities play a central role and where there is little reference to the compelling power of scientific evidence. Then, after a pause in his reply, he goes on to deny the long–term significance of such factors and claims that scientific facts will finally be decisive. This disjunction between the description of the current state of affairs and the expectation that, in the end, the truth will out was so frequent that Gilbert and Mulkay came to refer to this interpretative step as the 'truth will out device' or TWOD. TWODs feature in these scientists' replies as a characterization of science which assists in handling potential strains in the institution of science between personal and technical considerations.

A second excerpt from this study allows the role of TWODs to be appreciated. This biochemist (B) tells the interviewer (I) (Gilbert and Mulkay, 1984, p. 96):

B: You haven't touched on personalities *very* much. Spencer and so on. I'm not sure I want to talk about them. But I think they *have* contributed.
(He then talks about some personal bitternesses in his area of science and concludes by saying)

B: I think there *is* a subjective element.

I: Do you have any idea how this personal element gets eliminated?

B: Only because a sufficient number of experimenters try to make the position clear ... Predictions will be followed up, more experiments done, and in the fullness of time a much clearer position will become apparent ... And then, any personal rivalry will be seen for what it was, in relation to the facts, as they become more fully established.

I: So the experimental evidence ...

B: At the end of the day solves everything (general laughter) [italics in original].

Thus, towards the end of this interview the biochemist explicitly comments that his interviewers have not really addressed the issue of personalities. He asserts that there *is* a personal element which influences the course of scientific debate and provides stories to confirm the importance of this element. When asked about the apparent conflict between the strong influence of personal factors and the technical rigour of science, B offers a TWOD. He asserts that 'in the fullness of time' the balance of evidence will become clear. His final claim that the evidence resolves everything 'at the end of the day' is so blithely optimistic and Panglossian that even he is moved to laughter.

In both these cases the TWOD is offered as a way out of an interpretative difficulty. The biochemists have both testified to the significance of contingent, personal factors in scientific choice and in the formation of scientific opinion yet they also wish to defend the legitimacy of science. That they are actually coping with a perceived difficulty is indicated by aspects of the interaction in these interviews: in the first case the scientist conspicuously paused before supplying the TWOD and the second biochemist's laughter indicated his discomfiture at his reply. TWODs thus provide a resource for coping with the diversity of scientific opinions whilst still maintaining that, *in the end*, there is one correct scientific viewpoint. The authority of science is protected against the potentially damaging influence of particular disagreements between practising scientists. In the sense outlined at the start of this section, TWODs operate as an internal ideology for science.

A contrasting example of an image of science which is mobilized to further scientists' interests outside their immediate community comes from another study by Mulkay (1976). This study took the form of a review of the sociological theory of science propounded by Robert Merton (1973). Merton was concerned to understand how imperfect and seldom disinterested human actors could carry out the project of science so successfully. He argued that a number of norms regulating scientific conduct must have been institutionalized to permit the sustained growth of scientific knowledge. Drawing evidence from the descriptions of science offered by illustrious scientists, from observations of scientific behaviour and from notorious cases of scientific controversy, Merton considered that it was possible to formulate a short series of norms which characterized science. Briefly, these norms – of universality, communality, disinterestedness and organized scepticism – dictated that scientists should treat all scientific claims on the basis of their merits; disseminate their findings freely; not

48

seek to capitalize unfairly on their results; and suspend judgement on complex issues until all the relevant information was to hand.

These normative requirements may sometimes conflict with scientists' immediate self-interest or ambitions. Such a case is readily imaginable for the norm of communality; scientists may not wish to pass on their knowledge or new findings immediately, since they may prefer to develop the work to enhance their own fame or uniqueness. Merton argues that the normative structure operates to contain such temptations, partly because scientists internalize the norms and come to accept them as morally correct and partly because the norms are supported by social sanctions. Evidence for this viewpoint derives, as has been mentioned, from the statements of renowned scientists and also from the responses of scientists to those instances where the norms are disregarded. An example of this is supplied by the case, discussed by Merton (1973, p. 274), of the eighteenth–century scientist Henry Cavendish who, although brilliant, did not trouble to pass on his results but experimented for the love of it. Cavendish, Merton claims, not only failed to receive the praise and honour merited by his endeavours but attracted criticism from other esteemed scientists for not offering his work to the community. A negative reaction to an infraction of the norm (in this case, communality) is seen as evidence for the norm's general validity.

Mulkay (1976) noted that this account of the social functioning of science faced considerable difficulties. For example, scientists are apparently rewarded for producing published scientific results which others find useful in their work, not for adherence to the norms. Moreover, the norms are individualistic in flavour and seem to apply best to the kind of science practised in the nineteenth century and early in this century rather than to contemporary science which, as will be seen in the next chapter, demands large teams of researchers. Finally the norms make sense only if one has a very empiricistic notion of the scientific endeavour, believing that scientific knowledge stems more or less directly from close observation and experiment. As we saw in the last chapter, scientific theories often have to be retained in the light of apparently conflicting evidence. There is little basis for organized scepticism and disinterestedness if skill and experience have to be used in passing scientific judgements.

In the face of these problems Mulkay suggested that, in the norms, Merton had codified one frequently used but partial image of science. Mulkay notes that under other circumstances scientists will invoke characteristics of science which appear to stand in contradiction to the norms; they will talk for instance of the need for commitment or of the value of keeping one's findings to oneself until one can feel confident of their significance. For this reason Mulkay prefers to view the norms not as a straightforward depiction of scientific conduct but as a formal statement of a moral image of scientific behaviour which circulates within the scientific community. On this interpretation, the normative analysis is not actually false. It does however falsely diagnose the status of the normative vocabulary. For Mulkay (1976, p. 645):

> the standardized verbal formulations to be found in the scientific community provide a repertoire which can be used flexibly to categorize professional actions differently in various social contexts and, presumably, in accordance with varying social interests.

The author cautiously suggests that the normative vocabularies may be of utility to scientists in pursuing their social objectives. He provides two brief examples to illustrate this possibility.

On the one hand the vocabularies can be mobilized in the course of disputes between scientists. In this regard the case of competition between two British centres for radioastronomy is cited. One group, based in Cambridge, had results indicating a new source of radiation from space. When the results were formally published the other group alleged that there had been an undue delay in making the information available. This accusation was couched in terms of the unscientific nature of the Cambridge group's conduct. It was argued that the properly scientific thing to have done was to have passed on the information as soon as it became available. The 'offending' group were able to reply that the data were so important and potentially sensational that it was only responsible of them to substantiate the claim carefully before publication and to provide some sort of theoretical background lest the nature of the evidence be misunderstood. A dispute about the distribution of work opportunites was mediated through contrasting images of correct scientific conduct.

The second arena for the application of this moral vocabularly proposed by Mulkay is in relation to scientists' dealings with outsiders such as governments and official agencies. In a brief overview of the possible links, Mulkay draws on studies of the politics of the United States scientific community. In this context scientists' ability to present the scientific community as disinterested and self–regulating was important since it allowed them to argue against direct governmental scrutiny of the scientific research system and justified the largely autonomous running of science. Scientists could argue that the inherently moral nature of science made such scrutiny unnecessary. Furthermore, science could actually take a form of moral lead. The open, impartial and universalistic characterization of the scientific enterprise could be used to claim, in Tobey's (1971, p. 13) words, that 'American democracy is the political version of the scientific method'. The values of science, supposedly arising from the very nature of scientific activity, could even be employed to offer a model to the political community. Science represented the ideal democracy.

Putting the Ideology to Work

The suggestion that the image of science offers an ideology which scientific actors can use in pursuing their interests within professional disputes or with outside agencies is put forward only briefly by Mulkay. Recently Gieryn (1983) has sought to document further cases of the use of what he terms 'professional ideologies' of science. One particular use of such ideologies to which he draws attention is to erect divisions between science and other competing institutional bases of knowledge. One case which he discusses exhibits this process particularly well. Gieryn examines the work of the nineteenth–century scientist John Tyndall. Tyndall was employed as a scientific officer and teacher at the Royal Institution in London. This organization was originally founded to encourage the spread of scientific understanding and to foster the utilization of scientific knowledge for practical purposes (Berman, 1978). The lecture audience was composed of members of the public rather than of students in the regular sense. In the mid–nineteenth century the research or career opportunities available to British scientists were very restricted, and, according to Gieryn (1983, p. 784), Tyndall 'used his visible position at the Royal Institution to promote a variety of ideological arguments to justify scientists' requests for greater public support'.

Specifically, Tyndall argued in favour of science by trying to show how science differed from, and was superior to, other types of claim to knowledge which were seen to have overlapping authority. As I mentioned in the Introduction, even early in the nineteenth century science was coming to be seen by senior members of the Anglican Church as a formidable way of authenticating claims to knowledge. Although science and religion frequently managed to coexist quite amicably, and although many scientists had pronounced religious convictions, there was always the possibility that scientists' claims would conflict with established religious beliefs. Such a case, of course, was provided by the debates which raged in the 1860s over Darwin's *The Origin of Species*. Tyndall selected cases such as this in which the systems of knowledge did come into direct competition in order to make the case for science more insistently.

The issues at stake were far from wholly intellectual. The church still maintained enormous influence over the shaping of education so that the role to be played by science in the school and university curriculum had to be contested. Moreover, the clerical profession retained great cultural and social prestige. The possibility of scientists and a scientific worldview rising to cultural authority was hindered by clerical pre–eminence. In mounting his opposition to the church's intellectual authority Tyndall chose some direct techniques of opposition. Gieryn records (1983, p. 785) that Tyndall proposed to test the power of prayer in an experimental way by comparing indices of health in a hospital being prayed for with similar indicators for the situation when the organized prayer was absent. But Tyndall also sought to spell out the nature of science in many lectures and addresses by contrasting science with religion. He picked out a series of distinguishing features which were intended to reveal the contrast. Thus science is empirical and verifiable, whereas religion appeals to mysterious and supernatural entities; science is of practical utility, whereas religion is, at best, of value as an emotional support; where religion is dogmatic, science is sceptical; and science is objective and distinterested, while religion has a subjective attraction. To suggest the spirit of these arguments, Gieryn cites Tyndall's (1905, p. 307) claim that 'The first condition of success [in science] is patient industry, an honest receptivity, and a willingness to abandon all preconceived notions, however cherished, if they be found to contradict the truth.' To maximize the contrast with the dogmatism of religious belief, Tyndall stresses the detachment of scientists and their lack of commitment to any particular theoretical system. Yet as we have seen both in this chapter and in the preceding one, it is not possible and probably not even desirable for scientists to 'abandon all preconceived notions'.

The intellectual authority of the church was not the only obstacle faced by Tyndall. Gieryn records that 'Victorian mechanicians and engineers' also provided a challenge. The difficulty here operated on a number of levels. For one thing, as engineers themselves began to professionalize they aimed to build up intellectual monopolies in areas, such as advising on public works like the establishment of lighthouses, where scientists also wished to work. Some prominent practical men were also inclined to doubt the actual utility of science, believing it to be too abstract and liable to hinder technical progress by seducing those with engineering skills away into academic pursuits. Worse still from a position like Tyndall's, the strong claims for practical utility which scientists were obliged to make to compete in this arena had the ironical consequence of handicapping them elsewhere. Thus, for example, the traditional universities would be even more reluctant to establish or enlarge science departments if science was thought to be merely a practical undertaking.

Tyndall accordingly sought to elaborate a view of science which established its special practical value whilst, at the same time, indicating that science represented a venerable form of learning. The central ingredient of this image of science was its ability to penetrate to the causes of natural processes, an ability which science owed to its theoretical component. Thus he described engineering as a form of knowledge fundamentally built on trial and error and contrasted this with the systematic experimental method of science. Such a foundation meant that science (1905, pp. 95–6) 'would not be worthy of its name and fame if it halted at facts, however practically useful, and neglected the laws which accompany and rule the phenomena'. The inherent superiority of science is indicated because only science permits access to the causal laws which 'rule' the world of facts.

Gieryn records how Tyndall was able to supplement this basic division between science and engineering through, for example, an unfavourable contrast between science as a labour of love and engineering as the pursuit of business. Indeed, Tyndall suggests, the profit motive among engineers may actually lead them to compromise their intellectual judgement, since there are material incentives to promote one's own designs, plans, or processes even if they are not absolutely the best. Scientists, of course, can be expected to be impartial and disinterested. The final aspect of this portrait of science and engineering is the one which allows science to establish a clear superiority; for, in addition to its technical value, science 'has nobler uses as a means of intellectual discipline and as the epitome of human culture' (Gieryn 1983, p. 787). In Tyndall's own words (1905, p. 101):

The people who demand of science practical uses forget, or do not know, that it also is great as a means of culture – that the knowledge of this wonderful universe is a thing profitable in itself, and requiring no practical application to justify its pursuit.

Knowledge of this character would, naturally, deserve a place among the finest intellectual disciplines available in a university.

In this way Tyndall skilfully mobilizes an image of the nature of scientific activity and scientific knowledge which adds to the public status of science. He does this, as Gieryn points out, by demarcating science from other, possibly rival intellectual institutions. The image or professional ideology of science simultaneously stresses its positive characteristics and reveals the weaknesses of its rivals. The factor which allows Gieryn with some confidence to treat this image of science as a professional ideology rather than as a correct depiction of the essence of science is the variation in the image from one context to another. Just as we saw that participants in a dispute over scientific priority gave different emphasis to the importance of communality and the free exchange of all information, so Tyndall placed a different stress on the constitutive features of scientific inquiry at different times. As described above, in making the contrast with religious authority Tyndall emphasized the factual nature of scientific knowledge. To develop the contrast with the dogmatic assumptions of religious systems he explicitly claimed that, in science, theory was subordinated to observation. Religion was hamstrung by its dealings with unseen and inscrutable phenomena. But when we turn to the comparison of science with engineering we find Tyndall claiming (1905, p. 80) that 'One of the most important functions of physical science ... is to enable us by means of the sensible processes of Nature to apprehend the insensible.' On such a view, knowledge of natural laws and the underlying causal processes takes science far above the merely empirical. The two sets of assertions place very different premiums on the significance of theory and of observation. Different images seem to have been produced to enable Tyndall to achieve rhetorical successes. As Gieryn (1983, p. 787) summarizes his claim:

> Tyndall demarcated science from these two obstacles [to the growth of scientific authority], but the characteristics attributed to science were different for each boundary ... Alternative repertoires were available for Tyndall's ideological self-descriptions of scientists: selection of one repertoire was

apparently guided by its effectiveness in constructing a boundary that rationalized scientists' requests for enlarged authority and public support.

Images of Knowledge in the Reproduction of Science

So far we have seen an image of science used in making sense of the diversity of scientific opinions; in moral arguments over scientific conduct; in the cultivation of a favourable opinion towards science among outsiders; and in the course of struggles with other institutions over intellectual authority. One further area in which images of the nature of science are brought into active use is scientific education. A recent case study concerning the geological teaching provided by Humphry Davy at the beginning of the nineteenth century, again at the Royal Institution, illustrates this usage (Yearley, 1985). Davy's lectures began with an introduction to the nature of science and, specifically, of geology. In these introductory statements he espoused a strongly empiricistic interpretation of science. In a similar manner to Tyndall, he stressed the absolute reliance of science on experiment and observation. Thus, he attributed advances in geology to such factors as '"observation and experiment", to "devotion to experiment", or "an undefatigable [*sic*] spirit of investigation"' (Yearley, 1985, p. 86). And he declared that 'attainment of the knowledge belonging to these highly interesting subjects is *founded almost wholly upon observation*' (italics added). Further support for the exclusive dependence of science on observation was available from his lecturing procedures. He presented his audience with rock samples or with painted landscapes from which to observe rock formations. In acting in this way he not only taught them about geology but imparted lessons about how geological knowledge should be acquired. The clear implication of such teaching strategies, when applied to beginners in the subject, is that significant advances in learning can be made through simple personal observation.

Against this orientation, another component of Davy's lectures was an account of the historical development of geological knowledge. In these cases:

> when Davy emphasises the distinctness of modern geological views he refers to '[former] *systems* and ... doctrines relating to nature [which], when compared with the truth and *theories*

of modern time, appear as the vain toys and amusements of children when contrasted with the useful occupations and pursuits of men'.

(Yearley, 1985, p. 86, italics added)

What is picked out as distinctive of the modern state of geological knowledge is the advanced quality of the *theoretical* understanding. Yet the introduction to the lectures contained virtually no mention of the importance of theory since it insisted that geology was paramountly observational.

The same duality is revealed in Davy's discussion of particular historical contributions to the advancement of geology. Thus on the one hand he attributes great importance to the empiricistic views of the philosopher Bacon to the effect that 'all the sciences could be nothing more than expressions of *facts*' (Yearley, 1985, p. 89, italics added). At the same time he reserves special note and praise for figures whose major contributions he describes in terms of *theoretical* advances which, in turn, he ascribes to their individual genius. Indeed, when referring to the influential work of individuals such as Robert Hooke, an English scientist of around a century before, Davy makes no reference to observation at all. The contribution of such people is presented as occurring at a level above mere observation. The result drawn from this case study of Davy's use of images of science in his lectures is summarized as follows (Yearley, 1985, p. 89): 'The two factors invoked to account for the most stupendous and crucial aspects of geology's history (theoretical knowledge and genius) are absent from the factual/observational science proclaimed in [Davy's] introduction.'

One interpretation of this dual image of science might be that Davy was in error; perhaps he was confused or was simply a poor historian or speaker. Yet he was a hugely successful public scientist. His lectures, given from 1805 to 1811, were enormously popular. No contemporary commentators attacked the presentation as inconsistent, and there is no reason to suppose that the lectures were perceived as inadequate. Furthermore, examination of his formal geological writings fails to disclose any such duality there. It therefore seems that this apparent inconstancy is a systematic feature of his public, pedagogical presentations of science. The conclusion of the case study is that the two images of science (one observational and implying that science is readily learned, and the other suggesting that advances in science are attributable to the work of genius) operate together to resolve some of the practical difficulties of teaching science. By emphasizing certain

56

alleged characteristics of science, Davy the lecturer is able to guarantee its accessibility and to reassure members of the audience that they can really benefit from a short lecture course. Equally, through the emphasis on theory and super–observational qualities, he is able to stress the value and special qualities of scientific knowledge and to show how superior modern knowledge is. In this way Davy employs alternative images of science to enlist the audience's interest in science and to socialize beginners into an acceptance of the scientific view of the earth. Once again, images of science are used flexibly in promoting the needs of the social movement of science.

Institutional Forms of the Social Movement of Science

Up to this point the focus of this chapter has been almost exclusively on intellectual and ideological aspects of science as a social movement. We have seen that scientists have been able to mobilize the considerable intellectual resources of science for various purposes such as the defence of science and the enlargement of its authority. But successful social movements are not character-ized solely by robust and flexible ideological resources; they also demand institutional strength. Early on, many of the requirements of science at this level could be met through developing or sharing the institutional resources of other bodies. Thus, some scientists could gain employment, and aspects of a scientific training could be guaranteed, if the universities would agree to admit the sciences to the curriculum. This, as was described above, was one of Tyndall's objectives in the nineteenth century. By raising the prestige of scientific learning he hoped to strengthen the position of the sciences in the leading universities.

As regards opportunities for the pursuit of scientific invest-igations and the communication of information and techniques, scientists were relatively free to form organizations. From at least the time of the 'scientific revolution' (as it is commonly known) of the seventeenth century, natural and experimental philosophers formed themselves into discussion and experimen-tation groups. Because these early researchers were often wealthy and able to finance the experiments themselves, the organizations they formed needed no external support. As with all societies, members sometimes failed to agree over practical decisions; there were difficulties in ensuring that subscriptions were paid, and disagreements arose over the eligibility of people for membership. But all the time they were essentially private societies, operating on the basis of shared interests, overlapping networks of friendship and respect and personal recommendation, scientific organizations were institutionally unremarkable. It was only in so far as their activities might be regarded as posing a threat to established sources of intellectual authority that groups of scientists risked encountering opposition. As will be seen in the next section of this chapter, such opposition constituted something of a problem for the earliest major scientific organization in Britain, the Royal Society of London. Formally established in 1660, the society experienced some difficulty in developing a suitable niche in the intellectual environment. In the uncertain context of the early years of the Restoration its activities were construed by some as a threat to the universities, to the College of Physicians and to the newly re–established religious order (Syfret, 1950, p. 36). But even then this society had the independence which followed from being financially self–supporting. Its members even largely financed its publications.

The society became more properly royal in 1662 when it received its charter but this did not indicate that support was forthcoming from the monarch; for, as Wood reports (1980, p. 1):

> Royal patronage [gave] the Society prestige, and consequently members, but next to no financial support. Moreover, it was known that Charles II maintained a somewhat amused view of his official body of philosophers.

In any case, many of the originators were members of the nobility or titled gentry; at least 43 out of 125 by one count (Mulligan and Mulligan, 1981, p. 336). Some measure of royal interest was thus not remarkable; it could hardly be asserted that the society was supported by the state.

From the outset members of the Royal Society were keen to argue in public for the great practical importance of their scientific work. On at least some occasions they adopted a highly utilitarian rhetoric, and it is tempting to suppose that royal or state support was exchanged for the practical fruits. Yet it was precisely the uselessness of their research which attracted the critical attention of contemporary commentators and overall the claims for utility seem to have been much exaggerated. At this time the development of technical devices which might be of use to the state, for example an accurate means of establishing longitude at sea, were often encouraged by holding contests. Such competitions were every bit as likely to be won by skilled craftsmen and machine makers as by reputed scientists. In the case of longitude determination, the prize was captured by a skilled clock maker, John Harrison (Cardwell, 1972b, p. 74). And, as will be seen in Chapter 4, even the new devices set to work in Britain's Industrial Revolution around a century later were not designed by the leading scientists.

Overall, therefore, scientific societies did not acquire a special status; they operated much like groups we would now regard as societies for the practical arts or as cultural organizations. They might almost as well have been chess clubs. Although the potential for conflict with other sources of authority on the natural world did mean that they could occasionally be seen as suspect, they were generally far from being at odds with religion. Virtually throughout the eighteenth century, at least in Britain, scientific societies were generally culturally favourable to established religion, promoting, as they did, the study and appreciation of the Creator's works and discouraging, if only by taking up their members' time, participation in other, more obviously sinful pastimes.

However, the continued suitability of these organizational arrangements for the promotion of the interest of scientists was not guaranteed. Some pressure was being exerted on existing organizations by the increased numbers of people interested in science and by their growing specialization. Societies concerned with selective aspects of science like geology, astronomy or zoology started to be formed in Britain from the last quarter of the eighteenth century. The Royal Society, which aimed to represent all the sciences of the natural world, struggled against this fragmentation. This was particularly the case when a specialist society fastened on to a topical science and appeared likely to threaten the level of support for the older society, as seems to have occurred in the case of the Geological Society of London at the beginning of the nineteenth century (Rudwick, 1963, p. 332).

A second difficulty occurred alongside specialization. Even in these newer societies, which sprang from an amateur background, the prime qualification for membership was still involvement, willingness and a certain degree of knowledge. The issues at stake here are exhibited very clearly in the case of nineteenth–century chemistry discussed by Russell (1983). As with many other disciplines, chemists had formed themselves into a national scholarly association in 1841. This society 'provided them with a forum for discussion, a *Journal* in which to publish their researches, a Library, a house in London but little else' (1983, p. 226). The society was designed for the promotion and co–ordination of knowledge. However, as was mentioned above in the discussion of Tyndall, around this time scientists began to perceive opportunities for scientific consultancy. Some chemists began to use their membership (fellowship) of the Chemical Society as an indicator of their ability when advertising their services as analysts of water purity or of adulterations in food and medicine.

In many respects this was a reasonable procedure, but other members of the society objected. It was certainly the case that the society did not have the express function of authenticating the skills of its members. Its interests were in any case broader than the kinds of competence needed for most forms of chemical employment. Yet senior chemists realized that if the trustworthiness of some chemists could be certified this would enhance the reputation of their science and allow enlarged financial rewards and status to accrue to their colleagues. In short, the chemists realized that if they formed a professional body they would be well placed to capitalize on the available opportunities. Since the Chemical Society contained a high concentration of the country's leading chemists it was tempting to transform it into a professsional body; but, as Russell reports (1983, p. 230), 'After much heart searching and some legal advice the Chemical Society Council had decided that the Charter of the latter presented insuperable difficulties to some kind of select Institute within a larger Society.' A new body, the Institute of Chemistry, was founded in 1877. It sought to certify chemical practitioners, to publicize and lobby for their services and to promote the professional interest of chemists.

This incident demonstrates that there is no single best organizational form for science. Although there may be some general similarity between the activities of different scientists this does not mean that the 'essence' of science determines the form that scientific societies should take. Scientific organizations have diversified to follow scientists' increasingly specific interests and to facilitate the commercial use of their skills. Different views of science have been put forward to accompany these changes; science has been presented, among other things, as an adjunct to religious devotion, as a new utilitarian philosophy and as a professional competence. Institutional flexibility has accompanied the ideological resources (identified in the earlier sections of this chapter) available to the social movement of science.

The Historical Construction of Science

Even if it is accepted that scientists' professional ideology and organizational arrangements do not stem from the 'essence' of science, there might still be objections to seeing science as a social and historical movement gradually built up by its members as they sought authority and status. Opponents of this view often point to the origins of science. For many years historians of science asked questions about the 'discovery' of science at the time of the 'scientific revolution'. By highlighting the contrast between science and preceding ways of regarding the natural world, scholars generated a puzzle about the emergence of the scientific worldview. To their eyes it seemed that science (as practised by Newton or Hooke) appeared so suddenly at the end of the seventeenth century that it could not have been constructed. Science must have been waiting to be discovered all along. On this view, the leaders of the scientific movement did not invent science, rather they somehow came upon the scientific method and scientific worldview.

Striking support for this view seemed to come from the fact that early scientific societies, and in particular the Royal Society of London, appeared to have spelled out their scientific orientation and ambitions so precisely and in such 'modern' terms. How could one understand such clarity and precision if science were not a pre–given, unified thing which could be discovered? Thus, in the case of the Royal Society, Robert Hooke had drafted statutes for the society in 1663 and Thomas Sprat wrote an account of its methods and ambitions which appeared four years later. These statements stressed the importance of experimental knowledge and observation. More significantly Hooke asserted (quoted in Mendelsohn, 1977, p. 18) that 'This Society will not own any hypothesis, system, or doctrine ... (not meddling with Divinity, Metaphysics, Moralls, Politicks, Grammar, Rhetorick, or Logick).' To many historians this prescription has appeared marvellously prophetic of the value neutrality of science. The most obvious explanation of Hooke's apparent foresight was that he understood what science needed to be and what it needed to avoid. Hooke had apparently glimpsed the necessary cognitive ingredients of natural science. On this interpretation it is pointless to talk about the *construction* of science, since Hooke had already arrived at its essence very early in the scientific revolution.

Anyone who wishes to regard science as having been socially constructed must take this challenge to a constructionist view seriously. Recently, historians of science have begun to re–examine the scientific revolution and the character of the early Royal Society. Despite some differences of interpretation amongst these historians, two important strands appear in their arguments which assist the constructionist case. The first strand concerns the status of claims like the one by Hooke just cited. It is suggested that these claims were made essentially for apologetic purposes; that is, as a defensive plea about the harmlessness of the work of the Royal Society. These claims can be treated in much the same way as the statements by Tyndall. The second step depends on interpreting the values espoused by Sprat, Hooke and their supporters as the outcome of a negotiation over the character of science and not as a discovery of how science has to be.

As was mentioned in the short discussion of the Royal Society in the last section, it was much criticized in its early years, perhaps most colourfully by the physician Henry Stubbe, who by the mid–1660s regarded himself as a supporter of 'Church, Court and University' (Syfret, 1950, p. 24). As his exchange of criticism with supporters of the society progressed, Stubbe came to assert that the members of the Royal Society should

> now abate of their pride and censoriousness, and be satisfied that they are not necessary to the world, except one have an Occasion to send to the East Indies to know what grows in America, or to South Wales for an account of Nova Zembla, or the Countries subject to the North and South–Pole: If all History and Antiquity be to be affronted most impudently; if false relations concerning Salt Petre, Cider, Birch water etc. seem requisite ... or the Education and Religion of our native Country changed, there is some use for this Association.
> (Quoted in Syfret, 1950, p. 36)

The published defences of the Royal Society and of its principles should be considered in the context of criticisms of this sort. Thus, taking the example of Sprat's *History of the Royal Society* (a defensive 'history' written only seven years after the society's foundation), it has recently been claimed by Wood (1980) that Sprat's account of the society's activities was neither balanced nor even representative of the wide range of members' interests. Instead (1980, p. 1),

> by a combination of subtle misrepresentation and selective exposition, Sprat portrayed [the society's work as tending to] further the aims of social and ecclesiastical stability and material prosperity, essential for the Royal Society since its

continued existence depended upon the creation of a social basis for the institutionalized pursuit of natural philosophy.

On this interpretation, Sprat was chiefly concerned to outline an uncontentious status for the kind of knowledge his society was pursuing. In particular, he suggested that this knowledge was not going to conflict with the (newly restored) authority of the church and crown. Accordingly, natural philosophy was presented as strictly empirical and limited, as atheoretical and as likely to lead to practically useful knowledge. The claimed avoidance of theoretical reasoning assisted Sprat in steering a careful course away from both scepticism and dogmatism (Wood, 1980, p. 10). The former might have risked the danger of associating the society with atheism, whereas the latter would have been associated with the 'enthusiastic' religion which had been the hallmark of the preceding revolutionary decades and was now to be avoided. The broadcast image of science was thus not shaped by the needs of science nor even, on Wood's account, by the views of the majority of society members. It was an image calculated to merge well with the demands of Restoration England. The Royal Society could claim to meet the religious needs of the Restoration by opposing both atheism and enthusiasm; it could offer material benefits to the economy; and its members offered themselves as humble, painstaking experimenters – the very model of a good subject.

If the claims of Sprat and Hooke are seen as primarily rhetorical there still remains a question about the origin of this vocabulary for defending science. This issue has been addressed by Jacob. Jacob (1975, p. 169) is more inclined than Wood to regard the values espoused by Sprat as representative of values which many members of the Royal Society would have endorsed. His claim is that decision not to meddle 'with Divinity, Metaphysics' and so on was not a way of putting such issues aside; rather (1975, p. 155), 'these practices, which on the surface seemed to be aimed at preventing the Society from becoming involved in politics and religion, were in fact designed to help the Society accomplish its political and religious mission'.

Briefly expressed, Jacob's argument runs as follows. Despite the Restoration the state faced great religious difficulties. There were too many nonconformists (essentially dissenting Protestant sects) for a state religion to be successfully imposed. Thus the return of the monarch could not be accompanied by a return to a unified state Anglican Church. In Jacob's view, a faction associated with the natural philosopher Robert Boyle considered the appropriate solution to be one of limited toleration. Its supporters attempted to demonstrate that limited tolerance was likely to be beneficial to the church, to the gentry and to traders and merchants (1975, p. 158). For example, in the case of the church, toleration would at least give determined dissenters no cause to be hostile and might win some dissenters back. Toleration should also prevent further divides arising and weakening the established church. In the case of the Royal Society, where Boyle and his friends were strongly placed, the decision to avoid meddling with religion and its political upshots is a practical device to promote tolerance. It is in this sense that the policy of the Royal Society, far from being non–religious, was in Jacob's words designed to 'accomplish its political and religious mission'. The suspension of overt political and religious debate was actually expected to promote one particular politico–religious tendency.

The applicability of Jacob's analysis to the ethos of the whole of the early Royal Society has not gone unchallenged (see Wood, 1980, p. 4). But between Wood's account of the rhetorical role of the positivistic version of science espoused by the Royal Society's spokesmen and Jacob's claims about its political and religious function it is clear that the profile of science adopted by the Royal Society was moulded by social forces. The society's natural philosophers did not simply discover how science had to be; they devised a version of science which was viable in the specific and difficult context of the early Restoration. The constituent elements of science adopted by the social movement of science arose out of a process of social selection rather than from foresight about the necessary recipe for cognitive success (van den Daele, 1977).

Conclusion: the Ideological Utility of Science

Whereas in Chapter 1 the emphasis was on the socially negotiated character of particular scientific knowledge claims, the discussion in this chapter has focused on the historical construction of the authority of science *per se*. Science, as a way of gaining knowledge about the natural world, has developed as a social movement. Its constituent elements have been socially negotiated; so also has been its area of authority. Assisted by its ideological and institutional flexibility the social movement of science has achieved great success; as was recorded in the Introduction, by the nineteenth century science stood as a 'norm of truth' for 'cultured Victorians' (Cannon,

1978, p. 2). In Chapter 4 we shall come on to the issue of the extent to which the rise of science was driven by its technical utility and its appropriation by the state. But there remains one further reason for the general success of science as a form of intellectual authority: the suitability of science for ideological use. In this context ideology refers not to an image cultivated by a group or profession, but to ideas which are used to gain legitimacy. Ideas may be ideologically harnessed, for example, to support a political system or to justify social inequalities and discrimination. This meaning of ideology will be referred to as 'legitimatory ideology' to distinguish it from the professional ideologies discussed elsewhere in this chapter.

Science offers a potential advantage over other forms of intellectual authority in that it makes claims which are said to be founded in natural inevitability. It is one thing, for example, to say that wives ought to be subordinate to their husbands, as most Christian (and not only Christian) teaching has done. But if, in addition, it can be claimed that scientific evidence demonstrates that women are in some way naturally submissive or biologically inferior, this is stronger and in many respects more indubitable. Science, it can be said, offers the ability to 'naturalize' social, political, or cultural differences. For example, in the case of the phrenologists discussed at the end of the last chapter, the Edinburgh advocates of the new science of the brain tied phrenology into their political programme. If the reformists could show that people were predisposed by the structure of their brain to adopt certain stations in life they would have a naturalistic basis for their political programme. As Shapin (1979, p. 146) reported, the Edinburgh establishment which opposed the reforms also took to science to demonstrate that there was no natural basis for the reformers' proposals. In a similar way, studies by nineteenth–century British ethnologists like Mackintosh (1866, p. 16) disclosed that the native inhabitants of Ireland were 'deficient in application to deep study, but possessed of *great concentration in monotonous or purely mechanical occupations,* such as hop–picking, reaping, weaving, etc.' (italics in original). Such reports helped to naturalize Britain's control over its imperial territories and their inhabitants.

As these two examples indicate, scientific authority could be enlisted in support of social change or in the cause of the status quo. Either way, scientific evidence is powerful precisely because it adds the special legitimacy of making a social or political distinction appear naturally inevitable. Appeals to what is natural are not exclusively made through the medium of scientific authority, but scientific claims are a particularly effective resource in supporting

65

legitimatory ideologies. Sociologists often refer to the process of making differences appear to depend on natural facts as reification (Giddens, 1979, p. 195). In its history as a social movement science has frequently been sponsored because of the reificatory service it can perform. Overall, therefore, the historical success of the social movement of science has had a threefold basis: its conceptual vitality, the flexibility of its institutional and (professional) ideological resources, and its adaptability to a variety of legitimatory roles. Together these factors have permitted the great cultural authority of contemporary science to be assembled.

[3]
Scientific Work and the Research System

Introduction: the Shape of Scientific Work

Up to this point we have approached science chiefly as a form of knowledge. It has been examined as an undertaking aimed at the generation of understanding and authoritative knowledge of the natural world. But science is not simply a form of knowledge. As is shown by the elaborate work demanded by the experiments (described in Chapter 1) set up to study neutrinos and gravity waves, science demands refined skill and physical labour. For increasingly many people that labour provides a source of employment. This element of labour is often omitted from images and discussions of science. In the last chapter we looked at a number of images of science which scientists put forward from time to time. One recurrent image depicts scientists as working in a highly autonomous fashion. To the popular imagination such autonomy can often spill over into eccentricity. In its early stages scientific activity was indeed extraordinarily voluntary; scientists' work was largely self–directed, and the topics they researched were commonly freely chosen. People worked on science because, for whatever reason, they wished to. In moving to a description of present–day science it has to be recognized that the situation is now very different.

In the major industrial countries today science and technology provide work for hundreds of thousands of people. For an understanding of contemporary science and technology, and of the objectives and motivations of those involved in constructing science at present, we need to appreciate how science impinges on

67

them as work. The scientists who fit the stereotype of following their own interests and aiming at new knowledge constitute only a small fraction of the people employed in scientific work; by and large the only place where these conditions are even approximately met is in universities and related higher education and research institutions. The majority of scientists are occupied in applied science, in technology, as technical assistants, or as educators. An indication of the size of this majority is provided in Table 3.1.

From this table it can be seen that scientific, technical and related workers are very numerous. In Japan, for example, they total nearly three–quarters of a million; this is not very far below one per cent of the total population. However, the number of these workers employed in circumstances which approximate to the ideal of freely chosen and self-directed work is small. An attempt to indicate the size of this group is given in row (e) of the table. As mentioned above, the universities have generally been the setting in which scientists have had the most freedom to pursue work of their own choice. In the case of the European nations this group constitutes around one-tenth of the general scientific workforce. The data for the case of the United Kingdom do not include personnel employed in institutions of higher education so that the calculation cannot readily be made. But from other information[1] it can be calculated that university scientists and engineers numbered around 18,000 in 1978 (the year used in Table 3.1). This would give a value for the proportion of university researchers among all types of scientific personnel of approximately one in seventeen.

An alternative way of identifying the nature of the bulk of scientific and technical work is to examine the sources of funding for research and development work and to consider where that money is spent. For the case of the United Kingdom this information is provided in Table 3.2.

In each of the years covered by this table, research and development work in the universities and polytechnics has consumed around one-tenth of the total available funds. Whilst government is clearly the biggest overall supplier of funding (its contribution reaching nearly 50 per cent in each of these years), the majority of research and development is carried out in industry. This industrial research dwarfs the amount performed in the 'traditional' scientific surroundings of the higher education sector, outspending it in each year by a factor of about six to one. Furthermore, virtually as much research as is done in the higher education sector is carried out by the government itself in defence work; this is largely performed in specialized defence research establishments. Such work will be

Table 3.1 *Typical Distributions of Scientific and Technical Workers in Advanced Industrial Countries*

Country (date of figures)	USA (1983)	UK* (1978)	France (1979)	FRG (1981)	Hungary (1984)	Japan (1984)	Sweden (1983)
(a) Total scientists, engineers and technical and related personnel	n.a.	251,400	230,766	371,548	49,360	710,072	47,758
(b) Percentage of (a) employed in higher education	n.a.	n.a.	19.9	19.7	16.0	32.1	29.9
(c) Total scientists and engineers	728,600	86,500	72,889	128,162	22,518	531,612	19,081
(d) Percentage of (c) in higher education	13.7	n.a.	32.8	23.6	20.4	35.2	37.7
(e) Scientists and engineers in higher education as a percentage of (a)	n.a.	n.a.	10.4	8.2	9.3	26.4	15.1
(f) Scientists and engineers (c) per million population†	3,140	1,540	1,340	2,080	2,100	4,490	2,290

* The figures available for the UK exclude people occupied in higher education.
† The figures for population are those available in 1984, not in the year for which the other statistics were prepared. Any population changes in those years would have a negligible effect on the values of the ratios.
Sources: Adapted from UN, 1986, pp. 474–6; UNESCO, 1986, pp. V 37–V 40; *The Economist*, 1984, p. 8.

Table 3.2 *Support and Performance of UK Research and Development, Analysed by Sector**

Year	1972		1978		1983	
Sector	Amount of funding supplied	Proportion of work carried out	Amount of funding supplied	Proportion of work carried out	Amount of funding supplied	Proportion of work carried out
Government defence	22.1%	10.1%	25.4%	9.4%	25.7%	9.9%
Other government source	26.6%	15.5%	21.6%	12.2%	23.2%	12.8%
Universities, polytechnics and related institutions	1.0%	8.8%	0.8%	9.0%	0.4%	11.4%
Public corporations	5.8%	5.3%	7.4%	6.1%	43.6%†	63.2%
Private industry	37.7%	58.0%	36.8%	60.1%		
Other	6.8%	2.3%	8.0%	3.2%	7.1%	2.7%
Total cost	£1,313,400,000		£3,510,300,000		£6,583,000,000	

* The information in this table excludes expenditure on research in the social sciences.
† The figures for private industry and public corporations are not given separately for 1983.
Source: Figures from the UK Department of Trade and Industry published in Central Statistical Office, 1987, p. 221.

further discussed in Chapter 5. But even this represents only a part of the governmental expenditure on defence research, which in recent years has totalled over a quarter of the national research and development budget. The majority of defence and weapons research is thus undertaken for the government by private industry. (Additional defence work is undertaken by private industry on its own behalf.) It is not possible to move from these figures to statements about the numbers of researchers engaged in one sort of work or another. Defence research may, for example, be more capital intensive than biological research conducted in universities, so that universities may employ more people than the size of their proportion of expenditure implies. Furthermore, there are bound to be limitations to the statistics underlying these tables. The methods of counting and classification differ slightly from one country to another and all manner of everyday assumptions will have been built into the practical business of drawing up the figures. But, despite these shortcomings, it is apparent that there is good general agreement between the two tables indicating that only a small minority of scientists are employed as independent researchers – the kind who featured in the discussions of neutrino science and of gravity wave detection in Chapter 1.

Scientific Workers

Although they are in the great majority, scientific workers employed outside of the university and related sectors have received little attention from sociologists. Some studies of industrial scientists were conducted by analysts working with the image of science propounded by Merton (for example, Kornhauser, 1962) which suggested that science was characterized by an open and universalistic ethos. On the assumption that industrial research would not correspond to the demands of this ethos – commercial research tends to be secretive and to be directed by the demands of management or the market rather than by conceptions of inherent scientific interest – sociologists expected to find maladaptation and role conflict amongst scientists in industry. One detailed study, involving interviews with 283 scientists and technologists divided between those working in universities and those employed in industry followed up this suggestion. Ellis (1972, p. 204) looked at both the current perceptions and the expectations of industrial scientists and technologists. He found that scientists in industry did

71

not particularly prize such factors as 'Freedom to choose my own research projects' or 'The opportunity to pursue basic research in my field'. Indeed, in most respects the industrial scientific workers did not compare themselves at all unfavourably with people based in universities. Even in terms of the kind of research they were doing (Ellis, 1972, p. 195):

> Most of those interviewed expressed a definite preference for applied research; the reason most frequently given for this was that applied science provided more intrinsically satisfying results – a tangible end product – than those usually associated with basic or fundamental research.

Ellis's research did uncover some sources of dissatisfaction. Some respondents, for example, expressed discontent with the amount of time available for the pursuit of their own research interests. The chief drawback indicated by the industrial scientists (as opposed to technologists) in his survey was that they felt their abilities were under–used and believed that their skills might not be developed. This finding was reinforced in a study of science graduates entering commercial research. From his interviews Barnes (1971, pp. 161–3) reported that graduates generally adapted very readily to the commercial environment and soon exhibited identification with business values and priorities. Here again, dissatisfaction was recorded largely when graduates felt their expertise and authority were not properly recognized.

From the views expressed by Ellis and Barnes it might appear that the sources of dissatisfaction and frustration perceived by commercial scientific workers are comparatively slight and compensated by other aspects of their jobs. For example, Ellis's industrial respondents appeared reasonably satisfied with their salaries; much more so than their academic colleagues. But it is in the light of the dissatisfactions experienced by scientific employees that the work of Ellis and Barnes is in closest agreement with the claims put forward by several authors concerned with the role of scientists in modern capitalist societies. These latter authors, including Albury and Schwartz (1982), Gorz (1980) and Hales (1982), all emphasize the extent to which scientists in industry work on routinized tasks with little control over their work and little chance to develop their skills. For these authors the characteristic condition of scientific workers is one of alienation. In the words of Albury and Schwartz (1982, p. 117):

It would be exaggerating our case to say that scientific workers are the same as workers on, for example, a biscuit production line. Scientific work is more privileged with greater control over time, pace and hours than assembly–line work. But on the other hand, because of their reluctance to join trade unions, scientific workers are less protected than the technicians with whom they work and they are subject to speed up, night shift work, exposure to chemicals and radiation and forced redundancies, all of which have increased the negative aspects of the job over the last ten years.

Regrettably, these authors supply no systematic evidence to document their view that alienation is the increasingly typical lot of scientific workers in industry. They do, however, cite some examples which at least exemplify their claims. Thus they point to the routine nature of much scientific work and the relentless pressures for commercial justification of all research activities (Gorz, 1980, p. 275). Emphasis is also placed on the increased subdivision of complex scientific tasks which leads workers into ever narrower research (Hales, 1982, p. 61). This tendency, it is suggested, further decreases workers' autonomy, makes them more subject to the authority of scientific bosses, and takes away from them the opportunity to understand and appreciate their work as a whole. This latter point is, of course, directly analogous to the claim Marx made about one of the alienating consequences of the industrial division of labour.

These critical assertions about scientific work find some support from Roslender, who interviewed 114 scientific and technical workers in private and public industries in the UK. Of these people, fifty were in supervisory or managerial positions; the remainder performed exclusively technical work. Amongst the latter group Roslender (1987, pp. 8–9) found that 'Routinization appeared to be widespread and disheartening' and that 'Fragmentation of task also seemed widespread with the suggestion that this was a change from previous times'. A significant divide was apparent between the two groups of employees; the vast majority of those in supervisory and managerial positions claimed to experience extensive autonomy in their work and expressed satisfaction with their career prospects, whereas these views were held by only a minority of the technical workers (1987, p. 10). For Roslender the situation of most scientific and technical workers is becoming more like that of 'non–professional workers, most particularly manual workers' (1987, p. 13).

One conclusion that has been drawn from material of this sort is that, as scientific and technical work is proletarianized, scientific workers will come to identify politically with working–class interests (Albury and Schwartz, 1982, p. 116). It is even suggested that, because of their scientific education and training, such technical workers may constitute an intellectual 'fraction' of the working class. Claims of this kind are inevitably hard to assess. Often it is difficult to separate analysis from exhortation; Gorz's paper aims to convince scientific workers of their 'correct' class affiliation. More certainly, one can conclude that the majority of scientific and technical jobs will continue to be supplied by the commercial sector. And as more emphasis is placed on the competitive importance of innovation and technical capability (as will be seen in Chapter 4), the commercial pressures on scientific workers are likely to be intensified. It does not automatically follow from this that scientific work will be lacking in stimulation and enjoyment. However, so long as scientific training stresses the older image and ethos of autonomous research, workers are likely to develop false expectations and to face disaffection.

Science and Work in the Academic Sector

Notwithstanding their minority status, scientists in universities and similar institutions are the people chiefly represented in contemporary stereotypes of science and who provide the dominant image of the scientist. Science is customarily associated with the pursuit of novel knowledge of the natural world. Such scientific research, aimed at no immediate application or practical objective, is described as 'basic' research. Popular presentations of science, such as BBC Radio's *Science Now* broadcast, deal almost exclusively with science in this sense. But in the vast majority of cases even this relatively self–directed, basic science now has to be financed from sources beyond the wealth of the scientists themselves. A few field geologists, for example, may be able to get by without much special funding but even they depend for their salaries on a university or the state. Thus the great majority of basic science is supported in one way or another by the state and is not ultimately autonomous; it is, in Chubin's terms (1984, p. 5), 'kept science'. In the next chapter we shall look at what governments get, or hope to get, out of this arrangement in return for their sponsorship, but our present concern is the character of state–supported scientific work.

Up to this point a contrast has been drawn between scientific work in the commercial sector and that in universities. Yet developments internal to the latter sector mean that a good deal of university research is now quite unlike the individualistic, voluntary ideal. Most of these changes have been associated with the growth of what, after de Solla Price (1965), is now almost universally described as 'big science'. There are many aspects to the 'bigness' of modern science. The number of journals publishing science grows and grows; the number of professional meetings increases; and the number of scientific societies expands. But big science is not simply a magnification of little science; in two central areas big science is distinct from its precursor.

For one thing, some current scientific facilities are so expensive (that is, financially big) that there can be only very few in existence. Such a circumstance was mentioned in Chapter 1 where a vast subterranean tank was required to measure the flux of neutrinos. In some cases the expense is so great that researchers may have to share apparatus between a number of countries and agree to share time on a piece of machinery which needs to be in continuous use to make economic sense. A conspicuous example of this tendency would be the scientists working in the particle physics laboratories at CERN (the European Centre for Nuclear Research). A few scientists from each of the sponsoring countries get an opportunity to work on the vast accelerators buried beneath the Franco–Swiss border. Similar conditions apply in astronomy, where new observatories and telescopes cost enormous sums which have to be shared between scientists from many universities if not from several nations. Further instances would include ocean–going survey vessels and drilling rigs for the earth sciences. Perhaps the crowning example is provided by space science. In this case, however, the connection to military research is very intimate, as will be seen in Chapter 5. The impulse to share expenses is accordingly tempered by the need for secrecy.

The second specific feature of big science is the way in which it is conducted as team research. Very many people are required to run a large research installation such as CERN. Under these circumstances it is virtually impossible for researchers to claim any experiment as their own. The work is necessarily collaborative. Publications stemming from big science laboratories may have dozens of authors. The necessary teamwork means that the individualistic ethos of earlier scientific work is not appropriate to such collectivized research (Ziman, 1983). Thus, even where big science research is of a basic nature – that is, when the objective of the investigation is principally to satisfy our curiosity about the natural world, as with much astronomy – scientific work has come to resemble industrial scientific labour. Scientific research at CERN may not have any direct commercial aim but the scale of the enterprise tends to generate hierarchies and structures similar to those found in industrial laboratories. A thought experiment proposed by Ziman (1983, p. 4) captures this point well: 'Imagine that you are blindfolded, and driven round for several hours, and then released in a typical laboratory ... how long will it take you to decide where you are?' Ziman suggests it may take several days for the average observer to determine whether the laboratory to which he or she has been transported is in the commercial sector or not since the work practices, equipment and types of worker will be so similar.

It is for this reason that the authors discussed in the last section who stressed the alienating aspects of commercial scientific labour are also critical of basic research performed in universities and research centres. For them, teamwork usually means the acceptance of routine and hierarchically organized tasks, and collaboration means an advanced division of labour. In these circumstances, Albury and Schwartz suggest (1982, p. 116), it is the research managers who come off best. Their names feature prominently on the published papers even though they may do little of the routine work. Although such claims appear plausible there is little systematic evidence for them. An indication that research is regarded as a chore, by some scientific workers at least, rather than as personal project is available from a study published by Whitley. He notes (1978, p. 434) that in one particle physics laboratory studied it was claimed that 'because the civil service regarded research reports as a good measure of productivity, up to half of the total man hours available to the organization were spent [doing unimportant, unoriginal work] so that a suitable number of reports could be written'.

It is clearly impossible to make a general judgement about the extent of perceived dissatisfaction and alienation in big science. It can be stated, however, that much modern basic science is more routine, more open to bureaucratic supervision and more hierarchically organized than most commentators have supposed. But big science does not only have an impact on scientific work through the influence on the structure of labour. By drawing scientific inquiry into identifiable and costly research centres, big science makes the expenditure on science rise steeply and become more apparent. State support for scientific research facilities takes on the status of a major item in the budget. Accordingly, scientific research in the university and related sectors, or the research system, becomes the subject of overt political attention. Thus, in addition to the possible alienating aspects of big scientific work organization, the lives of scientists are affected by the politics of state support for science.

The Research System

The history of state support for scientific research will be briefly considered in the next chapter. At this point we are concerned with the consequences for the research system of the fact that it derives support from the state. These consequences have mostly only recently appeared. Of course early natural philosophers who were not themselves wealthy were dependent on patronage; what they worked on was thus open to some influence from their patrons (Hay, 1987). Equally, as Abir–Am (1982) has argued, in the years preceding the Second World War the Rockefeller Foundation exerted a great influence on the development of biological science by selectively providing funding for equipment for scientists who were willing to bring the techniques of physics to bear on biological materials. She concludes that the foundation employed 'a narrow technological definition of eligibility for support' which led to physicists 'colonizing' biological science (1982, p. 367). But, massive though this influence from the foundations was for many laboratories, it is only with the development of state support for the vast majority of basic scientific work that the problems of the research system become clear and practically universal.

It is currently the case that state governments routinely provide the bulk of funding for the pursuit of basic natural science. Exactly how this is organized varies from one country to another; the funds for science may come out of the education budget, out of the industry budget, or out of a special science budget. Whatever the arrangement in any particular country, the situation has many analytical similarities. The research system can be regarded as composed of a chain of decisions stretching from the part of government providing the finance down to the level of scientific workers employed in laboratories and research establishments. At one end the total sum available for science is stipulated in the course of the political process, while at the other end of the chain judgements about such matters as scientific publications are generally devolved entirely on to the scientific community. Scientists are largely responsible for rating and evaluating each other's work.

If these two types of decisions are regarded as the ends of the chain of decisions in the research system, one can then ask what the intervening decisions are and how they are made. Starting at the governmental end, the chain has approximately the following form. First there is the decision about how much to allocate to science as against other claims on the budget of the funding ministry. For example, if research is supported from the education budget, science spending will have to be balanced against claims like that for teachers' salaries. Next, it must be decided how much of the science sum to allocate to researchers working on basic science as against other forms of research. Such a decision may already be predetermined by the previous choice if, for instance, basic research is funded out of the industry budget. The sum set aside for basic science must then be apportioned in two senses. Choices have to be made about the amount of money to be devoted to different facets of research: research money for existing scientific personnel, backing for special research centres, the endowment of research fellowships and the provision of grants for research students. The second apportionment is between the competing basic sciences: for example, the disciplines of physics, chemistry, biology and so on. Even within these disciplines a selection may be needed between competing sub–disciplines or specialties; within geology, for instance, there are hard rock and soft rock divisions. Finally, the decision has to be made about how to allocate the funds available between the different university and college departments and the individual academics who apply for support in such a way as to ensure that good work is carried out, that skills are used and developed and that competence is maintained across a broad range of scientific areas. At some level, the ultimate goal of this series of choices is to promote the optimal production of good scientific knowledge. But, with the best will in the world, any set of choices made is bound to be contentious.

Whilst this whole chain of decisions has been referred to as the research *system*, it has not been planned or designed systematically. In most countries the chain has formed as the result of connecting up decisions which were coming to be made at a number of distinct levels. There is no single guiding procedure or theory which would allow the chain to be managed in an agreed best manner. Indeed, choices at opposing ends of the chain are made in radically differing ways. The general budget for science is allocated on a political basis, taking into account such factors as possible utilitarian benefits, the perceived political concern over science and accepted practice – the way the decision was arrived at last time or how other countries make the decision. At the foot of the chain, decisions are customarily made by scientists on the basis of the interest and value of their colleagues' work; work is assessed for its perceived scientific quality rather than in utilitarian terms. Since the separation between the ends of the chain is considerable the tension between the types of evaluation procedure may be tolerable. But a practical problem arises as the competing decision–making criteria spread towards the middle of the chain. Should one, for example, decide the allocation of financial support between scientific disciplines on a scientific, political, or utilitarian basis?

In approaching this issue it will prove helpful to examine one of the first explicit discussions of the bases for choice in science policy. As was mentioned above, up until the second half of this century scientific activity was funded from a variety of sources, and the funding was organized in an *ad hoc* manner. A review of the principles to be employed in deciding on the support for science could only be expected once the arrangements for scientific research had coalesced into a single research system. Thus, it was relatively soon after the coming together of the elements of a system in the United States that an eminent US physicist, Alvin Weinberg, attempted to clarify the basis for allocating support to science. He argued that criteria for decision making about science are essentially of two types (1963, p. 163); there are

> internal criteria and external criteria. Internal criteria are generated within the scientific field itself and answer the question: How well is the science done? External criteria are generated outside the scientific field and answer the question: Why pursue this particular science? Though both are important, I think the external criteria are the more important.

In these terms, support for expenditure on a particular scientific project could be proposed on an internal basis if, for example, it was believed that the scientific success was particularly likely. For Weinberg there are basically two internal criteria: '(1) Is the field ready for exploitation? (2) Are the scientists in the field really competent?' (1963, p. 163). Decisions about the internal criteria can be made, according to Weinberg, only by experts. Such assessments tend to be the ones with which scientists are

79

most at home. But since resources for science are not limitless, these criteria are very likely to be insufficient for determining what science should be afforded.

It is at this point that external considerations come into play. They direct policy–makers to justify choices about basic scientific research in terms of its likely material and cultural impact. According to Weinberg (1963, p. 164), 'Three external criteria can be recognized: technological merit, scientific merit and social merit.' Weinberg accepts that basic research cannot be assessed solely in terms of its likely utilitarian benefits; he accepts that to some degree scientific research has to be seen as an 'overhead' charged on the national economy. None the less, science planners may prefer one project over its rivals if that project is thought likely to lead to technical benefits which will serve the national economy. Social merit is assessed similarly although it is even harder to pin down. Weinberg himself (1963, p. 167) suggests that a social merit of some big science projects is that they may promote international co–operation and lessen national rivalries. The final criterion, scientific merit, is not unambiguously external. By scientific merit Weinberg means the likelihood that work in one area of science will have benefits for understanding in other areas. Such considerations may appear external to specialists engaged in an esoteric piece of research but will look distinctly internal to political eyes.

In one sense, however, the classification of the criteria does not matter too much. Weinberg is not claiming to offer new bases for scientific choice. Rather he is seeking to clarify the nature of the criteria which are generally called on, in the hope of encouraging better decision-making. Thus the allusion both to the cultural merits of science and to its technical benefits is familiar from the claims broadcast by Tyndall last century and discussed in Chapter 2. Weinberg's aim in setting out these five criteria is to help decision-makers be clear about the nature of the grounds to which they are appealing. The fact that he was unable to offer any way of weighting the various criteria against each other obviously limits the immediate practical utility of his work. None the less policy analysts have claimed that his statement of the relevant criteria has helped planners in coming to their decisions (Gibbons, 1970, p. 189), and the terms he employed have gained wide acceptance. For our purposes this consideration of the Weinberg criteria will assist in reviewing the bases of choice employed in the chain of decisions which underlies the research system.

Decision–Making in the Research System

If we return to the image of a chain of decisions it is clear that, at the governmental end, decisions about the total sum devoted to science are increasingly evaluated in utilitarian and economic terms. At this end the external criteria are paramount. Customarily, decisions at the other end of the chain have been made on internal grounds. There are, so to speak, both principled and sociological grounds for this. At the research end most decisions are made by some form of peer review. In the vitally important case of research proposals submitted by individuals or groups of scientists to funding bodies (in the UK, the Research Councils), the proposals are sorted by administrative staff but are put out for evaluation by other scientists. This procedure reflects Weinberg's claim that internal criteria can be assessed only by scientific experts. Other scientists with special knowledge of the field in which the proposal lies are asked to assess the scientific value of the project, its feasibility and the competence of the scientists to carry out the proposed work.

Such a system gives considerable autonomy to the scientific community; scientists are each other's judges. In Weinberg's terms, this autonomy would be defended in principle because it is only trained scientists who are in a position to evaluate the feasibility and importance of a project and who are able to gauge their fellows' skills. But the system also brings practical dividends. The potential for conflicts and rivalries between different disciplines or specialties can be avoided if, for the most part, each area is called on to make decisions only about its own projects

At the intermediate levels of the decision chain the basis for making assessments cannot be so easily classified. As scientific knowledge becomes increasingly specialized, scientists are less and less able to pass judgements on neighbouring areas of science on the basis of internal criteria. Someone with knowledge of one field will almost certainly not be competent to pass internal judgements on adjacent disciplines. Furthermore, it is not clear how different sciences can be compared. While it may just be possible to decide whether work on dinosaur extinction or on the earliest forms of life has the most to offer palaeontology, it is hard to imagine how the intrinsic scientific merit of these projects could be compared, for example, to the identification of new radio–wave emitting stars in deep space.

It is evident that such higher–level decisions are hard to make. But their difficulty is matched by their significance for the scientific workers in the research system. Thus, within the UK, one of the highest–level decisions concerns how the science budget is to be distributed between the Research Councils. There are five such bodies: the Science and Engineering Research Council, the Medical Research Council, the Natural Environment Research Council, the Economic and Social Research Council and the Agricultural and Food Research Council. It is unimaginably difficult to conceive how the division of money to these councils could be decided on grounds of scientific merit. Representatives from each of the bodies would claim that they support scientific work of the highest excellence. Their competition for funding is more likely to be pursued in the light of appeals to external criteria. The changes in the financial support for the Economic and Social Research Council (ESRC) between 1983–4 and 1984–5 can be seen in this light. During this time the council was under considerable public pressure and even had to change its name (it had formerly been known as the Social *Science* Research Council) because of claims, supported by the then secretary of state, that its work was not in a true sense scientific and not therefore as publicly beneficial as the work of the other councils (*Times Higher Education Supplement* [hereafter *THES*], 21 January 1983, p. 2). From the published figures for government support for the councils, it can be seen that the ESRC's proposed budget was the only one not to rise; indeed its proportion of the total Research Councils' proposed budget fell from 4.5 per cent to 4.1 per cent (the figure for 1981–2 had been 4.8 per cent: House of Commons, 1982, p. 2; 1984, p. 2; 1985, p. 2). According to a report by Paul Flather in the *THES* (21 January 1983, p. 2), funds from the social sciences 'had been diverted to help "new blood" in the natural science'. But apart from occasional changes in emphasis of this sort, an important consideration appears to be the continuation of pre–existing patterns of support. This apparent inertia is also evident further down the decision chain (Nelson, 1973). This tendency arises from a conservative strategy. Because none of the Research Councils nor, further down, any board within the councils wishes to give way to others, the strategy which causes least offence to all is to accept uniform treatment.

At these higher levels of decision–making it is clearly difficult to determine how to make the best choices. But the recent concern to limit public spending has brought out tensions even further down in the system. In a written answer provided to the UK Parliament on 12 March 1986 by George Walden, the under secretary of state at the Department of Education and Science, it was acknowledged that 'the increasing sophistication of scientific equipment and materials adds an average 10 per cent a year to their cost over and above average inflation' (House of Commons, 1986a, p. 471). Later in the answer it was stated that 'the Government's science budget has actually increased in real terms – by 6 per cent between 1981–82 and 1985–86 measured against average inflation'. Thus, while scientific inflation for this period may be calculated as approximately 46 per cent on top of average inflation, the increase in funding has been 6 per cent. The high rates of inflation in science are driven, among other things, by the demands for equipment of greater and greater technical specifications. No source of funding could be expected to meet such rates of annual increase indefinitely. When increases in state support do not match the rising costs, the competition between fields, and between the applicants in those fields, must be intensified. Given the rather conservative strategy likely to be adopted by the Research Councils in the face of falling real budgets, the area which we would expect to be most affected in the immediate term is the competition between scientists applying for research funds.

As was noted above, the customary procedure in basic research is that applications are submitted by individuals or groups of scientists proposing work to be carried out. Their applications are then assessed by other qualified persons in the same field. If the field is reasonably large and if the reviewer has no special vested interest in the outcome of the proposal then this system seems reasonable enough. Recently two shortcomings have arisen. The first is associated with the growth of big, collaborative research. The areas of research are no longer populated by large numbers of independent individuals; in big science they may well be restricted to a very small number of large research teams. If the team makes a joint proposal, as it normally would, the reviewer can only be a member of one of the other teams. Under these circumstances there is a danger of a lack of impartiality, since the team associated with the reviewer is likely to have its future funding prospects affected by the success or failure of the other group. Worse still, if funding is known to be stretched any reviewer, whether in a big or less big science, will know that his or her approval of a proposal will lessen his or her own chances of attracting funding, since the pool of available money will have shrunk. The anxieties associated with this problem were graphically expressed in a report in *The Guardian* (10 April 1986, p. 4) commenting on a paper published by a senior academic chemist (Williams, 1986). Williams's own comments addressed a related issue which we will come on to below but *The Guardian*'s columnist wrote: 'Scientific elder statesmen, who allocate much of the civil research budget, are feathering their scientific nests instead of awarding grants fairly, according to a senior Oxford chemist.'

The possibilities for pursuing-self interest to which this report refers are clearly present in the peer review system. These problems are accentuated when the system of peer review is used to administer a static or shrinking sum of money. The pursuit of self-interest does not even have to involve the rejection of competitors' applications; Irvine and Martin (1984a, p. 76) suggest that prominent representatives of the few large groups or centres in big sciences 'may reach a tacit agreement to support each other's grant and capital–equipment applications (regardless of what they might privately think of the merits of those applications), taking it in turns to seek funds.'

The management of the research system may show up further weaknesses when there are severe pressures on the amount of funding available. For one thing, as Collins (1983c, p. 7) has suggested, there may arise a tendency for peer reviewers to support work which has 'glamour without recklessness'. Reviewers will become too sensitive to the possibility of failure in research projects and may block highly innovative work. Equally, interdisciplinary work is likely to be discouraged because it is seen by reviewers or members of the boards of funding agencies as giving part of 'their' discipline's money to outsiders. Finally, an attempt may be made to preserve the funds available for the support of research scientists by making reductions elsewhere in the budget; this was the approach adopted by the ESRC, which responded to reduced funding by disproportionately cutting its provision of awards for research students (*THES*, 4 May 1984, p. 3, and 12 October 1984, p. 6).

Higher up the decision chain there is likely to be less flexibility; the conservative strategy of funding bodies facing budget cuts has already been mentioned. Indeed it is sometimes suggested that the scale of this inertia is indicated by the fact that the prevailing distribution of funding between the disciplines in British science is largely a historical residue from the postwar period which has never been overhauled (Irvine and Martin, 1984a, p. 72). In the UK, the Science and Engineering Research Council (SERC) receives around half of all the money available for the Research Councils. Leaving aside its support for engineering science, the SERC's costs are divided into five headings: there is an 'Astronomy, Space and Radio Board', a 'Nuclear Physics Board', two headings for administration and central activities

Table 3.3   *Proportion of Total Science and Engineering Research Council Science Expenditure Devoted to Big Science*[2]

Year	1981–2	1982–3	1983–4	1984–5
All sciences total (£000s)	145,717	158,912	168,476	180,172
Big science total (£000s)	81,073	88,762	95,374	105,604
Percentage on big science	55.6%	55.9%	56.6%	58.6%

Sources: Adapted from SERC, 1982, p. 11; 1983, p. 7; 1984, p. 7; 1985, p. 6.

and one 'Science Board' (SERC, 1985, p. 6). Expenditure on the work of the first two boards (space and nuclear research) may loosely be described as the commitment to big science. The proportion of the total financial support directed to big science is very large. The percentage of SERC science expenditure going on big science in recent years is shown in Table 3.3. As can be seen from the table, two areas of science consistently receive over half of all the funding. The remaining disciplines (chemistry, biology and parts of physics for example) are concentrated in one board.

When these figures are considered against the distribution of support two decades ago, it is clear that there has been a financial shift from nuclear physics to engineering and, to a lesser degree, to the science board (Department of Education and Science, 1985, p. 77). None the less, as Irvine and Martin (1984a, p. 72) comment, 'To those on the periphery of this structure, the system can all too easily appear to be little more than a means of ensuring that those funded generously in the past continue to be supported generously in the future.'

Concerns over CERN

We have already seen that there is a tendency for broad patterns of funding distribution between disciplines to be maintained. This is partly because there is neither a set of criteria nor any person with sufficient knowledge to make choices about the comparative importance of different basic sciences. But there is also the fact that any rearranging of the distribution would be extremely likely to provoke a hostile reaction. The potentially antagonistic logic of this situation was nicely captured in a feature by Jon Turney in the *THES* (Turney, 1983, p. 6); his article began: 'Are your experiments getting more expensive? Well, it's no good just asking the Goverment for more money. You need to ask for *some of the money they're giving to someone else*' (italics added).

An interesting candidate proposed for the role of victim in this reallocation of funds has been the subscription of the SERC to the particle physics facility at CERN. From the outset this European collaborative project was set to be expensive. Even in 1968 when a 300 GeV (giga–electron–volt) particle accelerator was proposed, a minority report was published alongside the majority's advice to support the project.[3] It argued that:

> We are convinced that to continue to spend such substantial sums in this direction is not in the national interest and that many scientists in this country, if they are properly appraised of the situation, will view the prospect with dismay. Can we expect them calmly to accept a slowing down of their activities whilst we go ahead with a huge project which will directly benefit only two or three hundred academic physicists?
>
> (Department of Education and Science, 1968, p. 55).

In 1983–4 the cost of the UK subscription to this research facility was around £32,131,000. This represented over 12.6 per cent of the whole SERC budget and nearly 6.5 per cent of the budget of all the Research Councils. (For 1984–5 this amount had risen to virtually 6.7 per cent and its value as a proportion of the SERC budget to over 12.7 per cent: House of Commons, 1984, p. 31; 1985, p. 30.) The cost was huge and rising and, because the pricing was carried out in Swiss francs, the UK subscription was liable to unpredictable fluctuations. In 1986–7 this last factor had a particularly marked effect (*The Independent*, 27 April 1987, p. 4).

CERN's considerable draw on the resources of the SERC attracted unfavourable attention from other scientists funded or hoping to be funded through that council. In response to this a review group was formed to examine the value and wisdom of the expenditure. Chaired by a crystallographer, Sir John Kendrew, the group submitted its report in June 1985 (Department of Education and Science, 1985). It was proposed that, after the completion of a building programme at CERN to which the

SERC was then committed, the UK should reduce its subscription payments. This should be achieved either through a negotiated reduction of all subscribers' payments (perhaps assisted by the introduction of new European participants) or, if such a move could not be agreed, through unilateral withdrawal. As would have been anticipated these proposals attracted strong opposition from physicists. They argued for the scientific merits of the work pursued at CERN and claimed that the costs were not as huge as at first appeared because the engineering and scientific work demanded by the construction and operation of the accelerators actually made a direct contribution to the UK economy. This contribution was composed both of orders and of incentives to innovative industry (see Schmied, 1977; and, for a recent popular account, *The Independent*, 11 May 1987, p. 13). There was further opposition on the grounds that any move towards withdrawal would be seen as proof of Britain's anti–European viewpoint.

By December 1985 it was reported that the UK 'under–secretary of state for higher education [was] treading carefully to avoid jeopardizing other agreements' (*THES*, 12 December 1985, p. 1). As a response to the Kendrew Report it was proposed that governmental spending on CERN should be reviewed. In February 1986 plans were announced to establish a 'European level committee' (*THES*, 28 February 1986, p. 1) to examine the report and consider European reactions. A committee selected by the president of the CERN council was asked in summer 1986 to respond to the Kendrew document within a year. The *THES* quotes the resolution of the council to the effect that the committee should report '"to the CERN council, and hence to the governments of the member states"'. The decision-making about this reorientation of science spending is thus likely (assuming that discussions go as planned) to have lasted over three years. At the time of writing, only an interim report has been produced. It is reported to be critical of weak management and to be offering 'long–term savings on staff costs [which, however,] do not meet the British target' (*THES*, 31 July 1987, p. 2).[4]

The history of these negotiations indicates just how complex and multi–layered the decisions about expenditure on scientific research are. In the end it may well be that the considerations about the level of commitment to CERN differ little from those which, according to Gibbons (1970, pp. 188–9), influenced some of the original decision-makers assessing the proposal for the 300 GeV accelerator. For them:

The main value to Britain appeared to lie in developing a feeling of confidence among the Common Market countries about her intentions to move 'towards Europe'. Establishing such a favourable climate of opinion could be considered a form of social benefit.

Although the Weinberg criteria provide a descriptive vocabulary for discussing the bases for policy choice, their use is clearly beset with the same problems as were the criteria for theory choice proposed by Kuhn and Newton–Smith. Supporters of each branch of science are likely to stress its great scientific value and are likely to perceive technical and social benefits which may flow from the pursuit of their research. None of the 'experts' whose advice may be sought are in a position to be disinterested. Worse still, it is unclear how a disinterested person could decide. In Chapter 1 we noted Kuhn's assertion (1977, p. 324) in relation to theory choice that 'two men fully committed to the same list of criteria of choice may nevertheless reach different conclusions'. The same difficulty inevitably arises for policy choice.

An attempt has recently been made to resolve some of these difficulties. Essentially the procedure has been to try to quantify aspects of the policy criteria. For example, in relation to the scientific benefit of expenditure on basic research one can look to measures of scientific productivity such as the number of papers produced, the frequency with which those papers are used and cited by other researchers and the esteem with which research is regarded by scientists in other countries or at other research facilities. Work of this sort has largely been pioneered by Irvine and Martin; they refer to this approach as the study of 'partial indicators of scientific progress' (Irvine and Martin, 1985, p. 300). These authors accept that no single measure could serve as an index of scientific quality; they have tried instead to assemble a collection of partial indicators from which an overall judgement can be made.

Critics have pointed to problems with any such procedure. Counting the number of papers produced, for example, does not take into account the quality of the science the paper contains. Irvine and Martin have accepted all such criticisms as indications of technical shortcomings – they agree that they have to be careful in their counting – but they insist that there is no better mechanism. However, the need for careful counting does mean that the technique works well only when directly comparable facilities are being examined. The cost effectiveness of one accelerator can persuasively be compared with that of another, but this has few implications for the broader questions facing a national scientific community. The calculations are also set entirely within the existing competitive framework of international science (Collins, 1985b). Some demands on the science budget arise more or less only because of this competitive pressure. There is a danger that in the course of calculations scientific merit will be reduced simply to competitive success.

The second proposal is that more accurate costings be made of the economic benefits of huge facilities like the CERN accelerators. Much of the doubt about the support for CERN was motivated by the great costs it imposed. The finance provided for those costs does not simply disappear. Costs arise because of the engineering work demanded by CERN; defenders of CERN point to the potential economic advantage that such work may bring to Europe (Krige and Pestre, 1985, p. 528). If this sort of consideration is going to be taken into account then an agreed method of accounting must be devised. And any such technique would be likely to work uncontentiously only in the case of comparing like with like. Furthermore, by placing such emphasis on the economic justifications for big science installations this approach runs the risk of overlooking their principal, scientific justification. Particle accelerators cannot be defended primarily as public works. As Schmied acknowledges (1977, p. 135):

It would be entirely mistaken to make a policy decision on basic science dependent on economic arguments or even to weight the utility created by research centers against returns on investments into other ventures such as, say, building a road.

Still, techniques such as these have great rhetorical power and more elaborate economic analysis and accounting can only be anticipated (Bianchi–Streit *et al.*, 1984).

Conclusion: Planned Research or Flexibility?

The final point to be taken up in this chapter concerns the nature of the connection between high–level policy decisions and the pursuit of scientific work. At the starkest level, it is clear that a decision to withdraw support from a big science facility would stop those scientists from carrying on their work. As the scrutiny of expenditure continues it appears that scientists will become more subject to policy–makers. One area where this effect is marked is in the competition between the Research Councils and between their boards. An avenue which they have all explored to some extent is that of pursuing increased external justification for their research activities. Their comparative position can be strengthened if they can claim to be supporting science which is simultaneously of the highest order and of public benefit. And it is rather easier to mount this kind of argument if the board or council can put its weight behind a package or programme than if it is simply responding to scientists' own proposals. Thus, if the MRC can state that it has a programme on AIDS or the ESRC an initiative on unemployment, they are able to make more robust claims than if they were simply to say they need millions of pounds in order to support diverse scientists. The flavour of this difficulty is neatly expressed in Figure 3.1.

Reproduced by permission of *New Scientist*

Figure 3.1

For this reason the policy bodies and the types of research they decide to support are becoming more important. Indeed this was the point which Williams was addressing when he reported (1986, p. 307) that the funding bodies have switched 'to the concept of special funding for "desirable" topics, to handing out money to "outstanding" groups, and to increases in direct funding to universities for "excellent" groups'. His fear is that, by directing research in this way, and particularly by bypassing peer review, the system will cease to be seen to be fair. He fears that 'Ability will be and has been substituted by self–interest.' Scientists will be drawn to where the cash is, and the cash is increasingly focused in research initiatives which are provided with an external justification or at least the semblance of one.

Williams's anxiety concerns the non–meritocratic consequences which follow from the direction of research. His anxieties would presumably be greater if it were the case that targeted research does not develop in the intended direction. The connection between the avowed aim of a proposal for research and the research actually undertaken is, by its very nature, open to interpretation and hard to pin down. But the suggestion that the connection is loose and flexible is supported by two studies with strongly contrasting methodologies.

The first of these studies (Farina and Gibbons, 1979; 1981) employed a statistical methodology. The authors set out to examine the implementation of policy decisions announced by the Science Research Council (SRC) (the precursor of the SERC). From time to time the SRC made statements concerning its aims in research funding and the approach it was going to adopt in distributing its funds. Farina and Gibbons undertook to compare such statements with the apparent implementation of these policies by examining the distribution and size of the grants allocated by the various SRC boards. In the period they studied (1965–74) the boards' view was that, since the costs of science were rising faster than inflation, some restriction on funding was necessary. Rather than penalize everyone equally, the planners aimed to let some regions and some topic areas take the main force of the cuts while others were encouraged. To retain some scientific activity at the forefront internationally, certain departments or individuals working in particular subjects would have to receive a disproportionate amount of funding. Besides the promotion of high quality work in areas of excellence, a further reason for encouraging an area was perceived national need – for example, for more engineering science.

However, when the authors took details of the financial commitments of the boards (the number of grants; the universities, departments and individuals to which grants were distributed; the average (median) grant size; and the distribution of the sizes of the grants) it became clear that there were a number of areas in which their statistical analyses appeared to run contrary to the stated plans of the SRC. For example, the proportion allocated in the form of engineering grants did not increase despite the avowed aim of encouraging this area. Moreover, the area of nuclear physics seemed to suffer no cutback despite the proposed reallocation of some of its financial support. There was also no evidence of the significant changes in levels of resource concentration which would have been anticipated. The authors suggest (1979, p. 319): 'it would appear that there is little correspondence between stated intentions and awarded grants when a board or committee's proportion of total SRC commitments is compared with stated funding priorities'. Their subsequent analyses addressed the question of grant size (Farina and Gibbons, 1981); here again they found little reflection in the figures of the stated intention of concentrating financial resources.

Farina and Gibbons do not opt for any single interpretation of these findings. But on any interpretation the results appear of clear importance for they indicate that funding support did not proceed in accordance with the Research Council's avowed policy. The explanation of this finding could be that the policy was just slow in coming into effect or that there was considerable inertia in the system of boards and committees. The high–level directives might simply have become dissipated in the course of the practical negotiations involved in the committees' allocation of particular awards. Yet an indication of the sociological implications of the finding can be gleaned from comments made in a later paper by one of these authors concerning an analysis of comparable Canadian data (Chapman and Farina, 1983). There the authors indicate that the implicit issue behind the analysis is (1983, pp. 320–1):

'What allocating mechanisms are at work which give rise to these trends?' ... The approach adopted in this study is to look for trends in the output data which impute [*sic*] certain characteristics about the allocation process. The imputed characteristics cannot be validated but, insomuch as they explain the observed trends presented earlier, they are at least internally consistent.

The authors' conviction is that the answer to this central question lies in the dynamics of funding committee meetings, in politicking and dealing between committees and in the operation of the peer review system. The important implication of this study for present purposes is that, with the complexity of the research system, decisions made at high levels can become diffused as they near implementation. There is more uncertainty and flexibility in the system than the statements of planners ever acknowledge. Even when the trend is towards unified programmes this effect can be expected to persist, since the kinds of goals around which programmes are built are themselves likely to be diffuse. Researchers will thus be able to present their work in such a way that it appears to conform to the goals of the programme whilst staying, in practice, quite close to their previous inclinations.

Further support for this suggestion about the flexibility of research plans comes from a detailed qualitative study of negotiations over funding proposals submitted by two US biologists. Myers followed the description and redescription of the proposed research work through successive drafts and after submission to the funding authorities. He recorded (1985, p. 228) how, after outlining his proposed work, one scientist finished 'his latest proposal with a paragraph on broadly suggested "spin–offs"'. His study reveals that scientists can routinely present their work as plausibly related to a number of diffuse and useful–sounding goals. Thus a move towards programmes with external justifications (such as the effects of unemployment) may serve to concentrate research moneys but will not necessarily produce different research or research of particular utility to that goal. In the light of such flexibility it is difficult to accept the strong claims about alienation in basic research introduced earlier. Since even attempts to plan research seem to leave the practising scientist with a great deal of discretion, one would not necessarily expect such attempts to be accompanied by a feeling of loss of control.

Science policy analysts have often ascribed apparent mismatches between planners' objectives and scientists' work to the inherently unpredictable nature of science. As Salomon (1973, p. 100) expresses it: 'In the case of science, however, the planning effort is subject to this special constraint: that it must set itself the *unforeseeable* as an object' (italics added). The two studies examined above indicate that there may be a further limitation: the exercise of scientists' ingenuity in devising interpretative connections between detailed scientific work and high level policy objectives Although the future prospect must be for further attempts at science planning and for research to be increasingly evaluated in terms of external, technological justifications, the effects of these developments on the working life of scientists remain little understood. Nor can it be taken for granted that this emphasis will result in the production of the sought–after golden eggs, as we shall see in the next chapter.

Notes: Chapter 3

1   The figures are from a chart, 'Guide to how Britain runs its science', which the journal *Nature* produces from time to time.

The chart used here, published in 1978, does not give a source for the official statistics on which it is based.

2 The figures used in this table exclude expenditure on postgraduate awards. They also tend slightly to underestimate the percentage of the science budget devoted to big science, since some of the administration costs, here included in the overall science total, should be assigned to engineering science and excluded from the chart. The overall science total should thus be a little lower. The big science budget would make up a larger percentage of this smaller total.

3 On the outcome of the plans for the 300 GeV accelerator, see Rose and Rose, 1970, pp. 233-9.

4 By late December 1987 no major decisions had apparently been made. A year's notice is required before withdrawing so Britain is now committed until the end of 1989. Some economies are expected in the light of the final report by the review committee but these may well be matched by the 'expansionist' plans of the new director–general who aims to acquire 'new accelerators to keep pace with developments in the United States' (*THES*, 18 December 1987, p. 2).

[4]

Science, Technology and Economic Success

Introduction: Innovation in the Economy

In the West we have lately grown accustomed to the continuous prospect of technical change. We remain unsurprised by great technical innovations. Innovations take a variety of forms. Sometimes products are offered which seem distinctive and new; the product itself appears as the outcome of technical innovation. Video cassette recorders, for example, appeared as a new form of consumer good. In the case of other items innovations are incorporated in the alleged process of continuous improvement; advertisements forever tell us that various brands of washing powder have been so improved that they now clean clothes better than they have ever been cleaned before. The same expectation holds for the world of work also; we are familiar with the idea that new methods and techniques of production will succeed our present machines and equipment. Innovation and the idea of technical improvement have become virtually synonymous with economic success. The purpose of this chapter is to review the role of scientific and technical knowledge in the generation of innovations and to examine the provision made by governments for such change. But to begin with it is necessary to be clear about the reasons why technical innovation is important.

The significance of technical change is assessed in a similar way by both Marxist and non–Marxist writers. Thus, according to liberal economic analysts, firms competing with each other in an open market will come to approximate a situation of 'perfect' competition. Under these circumstances the firms will end up offering essentially similar products at identical prices. In order to maintain a place in the market these firms are obliged to compete

95

by pricing their goods as cheaply as possible. To keep prices down they need to cut costs, and costs can be kept low by such measures as reducing employees' wages, purchasing raw materials more cheaply, or producing more efficiently. In this context technical innovations are generally important because they allow productivity to be increased. For example, a new manufacturing technique which allows the product to be assembled more quickly or with less wastage of raw materials is likely to confer a competitive advantage. Thus firms will be encouraged to seek out and adopt new technologies in order to continue to make a profit on their production. However, firms will also be competing to find and acquire the new technologies. A technology which is available to all producers will not result in any competitive advantage. Consequently, firms are inclined to try to buy selective rights to new technical improvements or to invest in their own research facility in order to have access to new technologies which their rivals do not have.

A technical innovation acquired in either of these ways may assist the firm by a change in the techniques used in production; if production can be speeded up or made more efficient the firm will be able to undercut its competitors whilst still making a profit. On the other hand the innovation may benefit the firm because it offers a new product over which the firm will temporarily hold a monopoly. Either way, technical change is important for ensuring adaptation in the market. Technical innovations are also regarded as holding longer–term consequences for the economy. Some economists (notably Schumpeter) have suggested that fundamental technical innovations underlie the long–term pattern of booms and contractions which has characterized the Western economy as a whole (Ronayne, 1984, p. 36). For example, the introduction of small internal combustion engines underlay the development of a whole generation of new products and new ways of doing things. In addition to the products which actually employed these engines (lorries, pumps, aircraft and so on) this innovation prompted other technical endeavours like the elaboration of road networks and the construction of suburban dwellings away from existing railway lines. On this view economic 'long waves' are governed by fundamental technical changes.

The Marxist position is basically similar. As is well known, Marx considered that he had identified crisis tendencies in the development of capitalist production. These trends were apparent at two levels. He suggested that the conditions of capitalist production would promote a political crisis as members of the working class came to perceive their common political interest in opposing the economic system. But he additionally claimed that the capitalist economy was heading towards exhaustion. This process, which for Marx underwrote the eventual political success of the working class, was expressed in terms of the law of the falling tendency of the rate of profit (Marx, 1971 [1894], pp. 200–52). Briefly expressed, Marx's argument was that because of competition between capitalist enterprises, successively greater amounts of profit would need to be reinvested in order to maintain up–to–date, competitive plant. Although such investment brought increased profits, a higher and higher proportion of those profits would have to be put aside for future investment (Sensat, 1979, pp. 59–62). Eventually, although one could not be sure exactly when, this consumption of the profits would become self–destructive. Marx's principal interest in this 'law' concerned its implications for the relationship between the capitalist and his or her employees. Marx believed that, in order to offset this tendency, capitalists would be obliged to turn to other ways of making economies; they would thus attempt to depress wage levels further and further, which would in turn incite the employees to political action.

This expectation now seems to have been frustrated. More recent Marxist analysts have generally seen the explanation for the postponement of the anticipated crisis in terms of technical innovation. On this view, the increases in productivity made available through technical innovations have been so vast as to offset any tendency for the rate of profit to decline. Marxists who adopt this interpretation now debate whether technical changes can thwart the 'law' indefinitely (Sensat, 1979, p. 59). Certainly many leading Marxist analysts such as Habermas adopt the view that the huge increases in productivity associated with technical change make the prospect of profound economic crises of this sort extremely unlikely. According to Habermas (1976, pp. 55–6):

> After the raising of *absolute* surplus value through physical force, lengthening the working day, recruiting underpaid labour forces (women, children), etc. had run up against natural boundaries (even in liberal capitalism, as the introduction of a normal working day shows), the raising of *relative* surplus value first took the form of utilizing *existing or externally generated* inventions and information for the development of the technical and human forces of production [italics in original].

Habermas concludes (1976, p. 57) 'that it is an empirical question whether the new form of production of surplus value can

97

compensate for the tendential fall in the rate of profit, that is, whether it can work against economic crisis'.

Although, in principle, it remains an open question whether technical change can continue to vitalize the economy, recent developments incline Habermas to expect continued vitality. In the period up to the early decades of this century industrialists were dependent on unplanned scientific and technical advances to suggest innovations. This is what Habermas means by the utilization of 'existing or externally generated inventions'. Such a system, in his view, still permitted the possibility of economic crises. However, in subsequent decades the state has stepped in to regulate the provision of scientific and technical skills and to provide the educational requirements needed. As Habermas expressed this in an interview (1974, p. 50):

> Now we have capital which is invested in the area of science, technology, education, and so on, in order to boost the productivity of labour ... the difference between late capitalism and liberal capitalism, or one relevant difference, is that we now incorporate these activities into the economic processes.

It is clear that from both of these economic standpoints (the liberal and the Marxist) technical change is central to the maintenance of the vitality of the modern economy. It is not simply that innovation leads to growth and wealth; innovation is indispensable if the economy is not to stagnate. As Freeman states (1974, p. 256), 'not to innovate is to die'.

The Institutionalization of Technical Change

Up to this point we have considered why technical change is of such major importance; less attention has been paid to the ways in which that change occurs. As the statements of Habermas imply, the amount of effort devoted to research and development, particularly for economic use, has increased vastly in the last century. But the change has not simply been an increase in scale. The increase can be said to have proceeded along three dimensions. There is first the issue highlighted by Habermas in the last passage cited. After at first depending on sources outside of industry altogether for technical change or on unplanned innovations devised during the course of work, industrialists began deliberately to cultivate 'research and development' (or 'R and D' as it is now known) in the expectation of technical advantage. Firms started to build up in–house research facilities or to commission independent authorities to do work for them. In Europe and North America the state too founded research facilities and began to provide scientific and technical training in order, in Habermas's words, 'to boost the productivity of labour'.

The second dimension of change is closely connected to the growth of large–scale manufacturing enterprises. There has long been a trend towards the concentration of the majority of production in the hands of fewer and fewer manufacturers. This was observed even at the end of the last century. By 1969, when Galbraith was writing of the new industrial state dominated by huge corporations, the 500 largest manufacturing corporations in the USA had 74 per cent of all the assets used in manufacturing (Galbraith, 1974, p. 89). The pervasiveness of these huge corporations holds strong implications for the development of technical research. In–house research and development facilities are not cheap. The simple cost of establishing an industrial laboratory may be crippling for a small firm. Such costs are much more easily met by a giant corporation. It is essentially for this reason that, as Freeman comments (1974, p. 199), 'the vast majority of small firms in OECD countries do not perform any organized research and development'.[1] Given the centrality of technical innovation to economic survival, the inability of small firms to afford much research has further disadvantaged them. As a general rule, therefore, large firms are better able to perform research and are thus more likely to succeed; they find themselves in a virtuous circle. According to Freeman (1974, p. 200), firms 'with more than 5,000 employees accounted for 89 per cent of all industrial R and D expenditures in the United States in 1970'. The figures for other industrialized countries are of a similar order.

This general association between a company's size and its commitment to research is not absolute. There are some small firms which specialize in technologically intensive products, and there are large firms with very restricted research and development efforts, for example in the food and beverage industry (George, 1977, p. 165). Moreover, Freeman reports that research intensity is far from directly related to company size (1974, p. 203); in many sectors of the economy the small firms which do undertake research spend as high a proportion of their resources on it as do their giant neighbours. Inevitably, however, they spend much less in absolute terms.

The third component of the institutionalization of research concerns the nature of the goal to which research is directed. According to Gorz (1976, p. 163) there has been a major historical shift in the objective of research expenditure: 'Up to the beginning of the Second World War by far the most important aim of research and technical innovation was to counteract by a *reduction in production costs* the tendency of the rate of profit to fall' [italics in original]. What Gorz has in mind here is that, at this time, innovations were primarily welcomed for their ability to increase the productivity of existing manufacturing and production processes. New machines and techniques were incorporated which allowed basically the same goods to be made more quickly, more cheaply, more reliably and with fewer workers producing the same quantity as before. This emphasis may be described as a concentration on process innovation.

Gorz contrasts this commercial strategy with a subsequent tactic which depended on innovating at the level of products (1976, p. 163):

> While innovations in production *processes* are still of decisive importance, they have developed *relatively* less rapidly, since the beginning of the 1950s, than innovations in the nature, style and presentation of consumer *products*. Instead of products evolving more slowly than the processes involved in their production, the situation is now often reversed [italics in original].

This change he associates with the potential for saturation of the available market with manufactured goods. On this view, once production efficiency has reached truly staggering levels consumers will not wish to acquire familiar goods as quickly as they can be made. The possibility of overproduction makes it attractive for manufacturers to look for new markets and new profits by introducing new product lines. Additionally, this trend away from process to product innovations coincides with a period of enormous growth for large corporations. While process innovations may hold many attractions for small firms operating in a market situation of nearly perfect competition, their appeal to large corporations is less intense. Contemporary large corporations commonly do not operate in anything like perfect competition; there are, for example, only a few producers in the high–volume car market in European countries. Accordingly, the primary interest of large businesses today is in the generation of innovative products. Their aim is to

100

introduce either new goods or new versions of existing products. Gorz argues (1976, p. 164) that such an objective is dictated by the requirement of continued profitability:

> There is only one way in which they can achieve this: by means of constant innovation in the field of consumer goods, whereby commodities for which the market is close to saturation–point are constantly made obsolete and are replaced by new, different, more sophisticated products with the same functions.

Under such circumstances, Gorz maintains, research and development are dedicated to 'accelerating the obsolescence and "moral depreciation" of commodities' and to replacing them with new products so as to 'create new opportunities for profitable investment of the growing mass of profits'.

Unfortunately, Gorz does not supply detailed information supporting his assertions nor any statistical figures to substantiate the extent of the change from process to product innovation. He also omits the role of state spending on research and development even though this accounts, as we saw in the last chapter, for around half of all such expenditure in many Western countries. And, since Gorz acknowledges that process innovations are still important, it is difficult to accept a strict interpretation of his claims about the changes in the type of innovations. Nevertheless it is relatively easy to identify good examples of modern commercial products which have many of the characteristics he highlights. If, for example, one thinks of the kinds of products which are commonly advertised in television commercials, some (albeit rather unsystematic) evidence is available. Successive versions of well-known brands of washing powder demonstrate the logic of this sort of innovation. Whilst each new generation of powdered detergent purports to offer unparalleled cleaning ability it is faced with the difficulty of outshining the last offering from the same manufacturer. If the cleaning power of the product was adequate, say, ten years ago, as the manufacturer no doubt claimed at the time, the improvements in the meantime have been to no great purpose. One cleaning product seems to have been 'replaced by [a] new, different, more sophisticated product with the same functions' in just the way Gorz outlines.

In one memorable series of advertisements for a new generation of a well–known washing powder which appeared on UK independent television, the presenters of the commercial built the appeal of the detergent around the scientific breakthrough on which it was allegedly based. White–coated dancers celebrated the technical triumph of the 'new system' product. This case offers good evidence that a definite product innovation was being offered. The new powder was not being advertised for its cheapness or economy; people were not being encouraged to use more washing powder; they were being offered a new, more sophisticated product which would replace their old washing powder.

In this case there is no data on the amount of research that went into the creation of the 'new system' detergent. And there is no reason not to believe that for certain kinds of washing task and under certain conditions the new powder outperforms its predecessor. This case is therefore no more than a suggestive illustration of the kinds of innovation on which Gorz passes his harsh judgement. In fact, Gorz (1976, p. 164) is extremely critical of the kind of innovation associated with new products launched by large companies. He claims:

> in the United States, and tendentially in Western Europe, monopolistic [by which he means 'large companies'] production is growing much more rapidly 'in value' than in terms of physical quantities. Monopolistic expansion relies less on increasing the volume of goods produced than on substituting for relatively simple goods more elaborate and costly goods the use–value of which is, however, no greater, and may even if fact be less.

His assertion here is that the innovative, sophisticated goods which are introduced are often of no more practical use – despite their sophistication – than the goods they replace. As a sociologist I am, of course, not competent to evaluate the use–value of soap powders, although to judge from past advertisers' rhetoric detergents have always had astonishing cleaning powers. But Gorz's principal point (despite appearances) is not to attack the quality of new products; rather his claim is that the decisive feature of innovations is that they benefit the producer. In particular, new products benefit the manufacturer by being more sophisticated and complex. The more sophisticated the product, the more work that will have had to be put into it; the more work that has been put into it, the higher its market value (what Gorz would call 'exchange–value') is likely to be.

An example from the food industry will help to make this point. Processed foods such as 'instant' mashed potatoes or potato waffles were product innovations. They are undoubtedly more technically sophisticated than simple potatoes but they are not obviously of much greater use–value. Of course, they might be said to have some 'use' advantages; they are generally simpler to prepare and easier to store than regular potatoes. Against this, their dietary value would probably not be thought to be as high. The exchange–value, on the other hand, is ordinarily much higher than that of potatoes. Gorz would regard these product innovations as ones which have been favoured because of their benefit to the food industry (for their potential for profitable trading) rather than for their value to the consumer.

At this point Gorz's claims are very close to a famous argument about 'producer sovereignty' advanced by Galbraith. Gorz puts forward a critical view of technical innovations on the grounds that they are designed more for the benefit of producers than for consumers. In order to make this criticism he employs the notion of use–value; in turn this notion demands that Gorz passes judgement on the usefulness of modern products. There is clearly a risk that Gorz's own views about what is desirable will enter into his criticism. For example, some consumers might claim that, for them, the convenience of 'instant' mashed potatoes makes them highly valuable. Galbraith adopts a slightly different tack; his contention is that modern industrial production is aimed at producing goods which, by and large, people have to be persuaded to want. Expressed briefly, his claim is that people have few basic needs; over and above these their wants are inherently subject to manipulation. Consequently, he argues that new products are typically supported with vast advertising budgets which create the demand for the product to satisfy; this strategy he describes as 'demand management' (1974, pp. 205–10). Such an interpretation implies that the notion of use–value is rather precarious. In Galbraith's view products do not have an objective use–value. If consumers can be persuaded that they would like to eat large numbers of potato waffles then they will in some sense have a use for those waffles even if other people maintain that waffles are an uneconomical, and possibly dietarily unsatisfactory, way of enjoying potatoes.

It should be noted that Galbraith is anxious not to exaggerate his claims. He acknowledges that producers are not fully sovereign; consumers cannot be persuaded to want just anything. But his view lends firm support to Gorz's suggestion that innovations may relate more to the interests of producers than to consumers. The significant point is that from Galbraith's reasoning one can arrive at this conclusion without having to make judgements about the true use–value of products.

A good example from a high–technology industry which will allow us to assess the plausibility of Gorz's criticisms of industrial innovations is supplied by Klass in his study of the pharmaceutical industry (1975). This industry seems to fit the characteristics stressed by Gorz very well. Pharmaceutical manufacture is research intensive. Klass reports (1975, p. 73) that in 1974 Hoffman La Roche spent more than $230 million on research and that this sum is estimated to constitute 15 per cent of the company's sales revenue. When it comes to the quality of innovations, Klass discusses the example of a drug backed by Burroughs Wellcome and supported by Hoffman La Roche (1975, pp. 100–12). This drug was an advanced antibacterial preparation which, the author shows, was advertised as effective for a great number of complaints. It was backed by a large amount of expensive literature, specialist conferences and considerable quantities of free samples. For an assessment of the value of the product Klass quotes the report of an independent professional journal, the *Medical Letter*, as follows (1975, p. 106):

> CONCLUSION: The combination of trimethoprim and sulfamethoxazole (Bactrim–Roche; Septra–Burroughs Wellcome) is now available for treatment of chronic urinary tract infection. This combination is effective in urinary tract infections, but in patients with diminished renal function other microbial drugs may be safer. One hundred tablets of Bactrim or Septra generally cost the patients between thirty and forty dollars, compared with about three dollars for generic sulfonamides.

In this case it appears that the sophistication of the new product, in terms of the research and development costs as well as the expenses of launching and publicizing it, is compensated by a rather high price which compares extremely unfavourably with the cost of similar generic drugs. When we turn to the utility of the drug it is said to be 'effective' for some complaints. But it was being advertised for use with other disorders for which, Klass argues (1975, p. 107), the utility was not agreed by medical experts. According to Klass, by aggressive marketing the drug was being offered for sale and application to cases to which it was probably unsuited and where it was of no greater utility than existing, cheaper preparations. This example persuasively bears out Gorz's claims; sales are encouraged to the advantage of the company even when it is far from clear that the benefit to patients has been increased. In this case, the specificity of the task which the drug is supposed to perform and the availability of an expert assessment allow us to talk with some confidence of use–value. To cite Gorz's remark again, pharmaceutical products (in this instance) 'are constantly made obsolete and are replaced by new, different, more sophisticated products with the same functions'. This example could even be regarded as exemplifying his apparently extreme claim, already quoted, that innovation sees the introduction of 'more elaborate and costly goods the use–value of which is, however, no greater, and *may even if fact be less*' (italics added).

Significant though this case appears, it must be accepted that it is not necessarily representative of all technical innovation since medical technology has some peculiar features. Doctors are not the principal consumers of the products they purchase from drug companies and in most countries they are not directly affected by the costs of the treatments they prescribe. They frequently have to deal with large numbers of patients and, for obvious ethical reasons, are not free to experiment for themselves with the efficacy of different treatments. They therefore tend to use only a small number of the range of available drugs. Companies have a great incentive to bring their products to doctors' attention and to stress the general applicability of their treatments in the hope of widespread and frequent use. In several respects this situation seems ideally adapted to the perpetuation of product innovations for which there may be little justification outside of commercial considerations.

There are two remaining issues which should be mentioned at this point although they will be developed further in the next chapter. The first is that little attention has been paid to the nature of technical improvement. Apart from the brief discussion of Galbraith's views, the idea that technical innovation produces goods which are technically better has been accepted more or less uncritically. Gorz's criticisms of modern products were based on the proposal that the goods are often not better for consumers. But even he did not demonstrate how betterness should be assessed. Detergent manufacturers claim that their products are getting better and better at cleaning; pharmaceutical companies claim that their drugs are getting more and more medically efficacious; Gorz suggests that new products are better at generating profits for the manufacturers. But they have not stopped to ask what better means. The difficulty here is very similar to that discussed in Chapter 1: all scientists are concerned to produce better knowledge but they often disagree over what better means. Even philosophers, such as Newton–Smith, who wished to accept that scientific knowledge definitely does get better, admitted that 'betterness' is by no means simple to assess; he proposed that betterness in science is evaluated along eight dimensions. There is no reason to be confident that technical betterness is any easier to decide. This issue of the meaning of technical improvement will be taken up in Chapter 5.

The second point is that manufacturers may be interested not only in technical innovations which bring profitability; other benefits may attract them. A useful illustrative case has recently been discussed by Winner (1985, p. 29). Drawing on a study by Ozanne (1967), he reports that in the mid-1880s pneumatic moulding machines were introduced in the foundry of an agricultural machinery manufacturing plant. This process innovation was introduced at a cost of around half a million dollars, but, according to Winner's account, the machines did not allow the reaping equipment to be produced any more cheaply or reliably. Although less–skilled employees were required for these moulding machines than for the previous method of manufacture, the new process cost more since the capital invested had been so high. Worse still, the new machines were unreliable. They were apparently dispensed with after three years. In this case, therefore, the innovative method cannot easily be seen as offering an improvement. Winner suggests that the machines did none the less offer an advantage; the advantage was precisely the machines' ability to displace the skilled labourers who, at this time, were responsible for militant trades union activity. The innovation was welcomed because it provided an improvement in the ability of the company controller to dominate his workforce. It is thus important to bear in mind the other possible benefits which technical change may be seen as conferring. Even if one accepts producer sovereignty, producers have interests other than immediate profitability.

Science and Technology Policy

Although technical change may be aimed at a variety of objectives, economic competition tends to concentrate effort on the reduction of production costs, the maintenance of market share and the ability to offer new styles and products. Manufacturers' profitability is influenced by changes in the sources and price of raw materials and by design and fashion, but the potential significance of technical innovations is enormous. In the international competition which has intensified between the Western economies, national governments have become increasingly concerned about the research and development effort being undertaken in their countries. In the last chapter we looked at what governments have done for science in sponsoring scientists' activities; science and technology policy concerns the pay–off which governments expect from this support.

Since the end of the 1970s there has been a pronounced tendency for governments to be more and more explicit about the utilitarian rationale for their support of expenditure on science. This trend has led to increased interest in the formal, rational arguments which can be offered for and against governments' funding of science. These arguments have focused on the question of whether science has special characteristics which mean that its economic benefits need to be carefully unlocked. Right from the initial institutionalization of natural science in the West, claims have been made about its utility. For long periods, however, these claims could readily be doubted. The impact of specifically scientific knowledge on the Industrial Revolution in England was slight; the men responsible for the major innovations were generally not favoured with a formal scientific education, and there was in any case only very limited teaching in 'scientific' subjects at England's two universities (Cardwell, 1972a, pp. 18–20). Admittedly, mathematics gained some standing as a subject of instruction in Cambridge in the eighteenth century following the success of Newton but it was more likely to be justified as a mental training for legal argument than for its virtue when applied to mechanical problems. The sole professional education which entailed the provision of some scientific instruction was medicine and those who were trained in this field had little incentive to look to manufacture as a means of livelihood. Thus, although the development of modern science and the growth of industry were associated in time, the actual connections were far fewer than we would now anticipate (for the case of mining and geology, for example, see Porter, 1973).

Connections between particular sciences and certain areas of manufacturing expertise began to emerge clearly in the nineteenth century, particularly in relation to chemical materials and preparations which did not occur naturally. However, the relationship between scientific expertise and economic success was no stronger than the correlation of success with other factors such as the availability of natural advantages (in terms of the presence of natural resources and the proximity of other industry), the existence of a skilled workforce and ordinary inventiveness. While some scientists (like Tyndall) and the spokespersons of professional bodies extolled the dramatic importance of science for the development of national wealth, it was possible for contemporaries to regard this as special pleading. Other factors could reasonably be held to be every bit as important as national scientific capacity, and governments, faced with the political pressures of office, could always delay the occasion when something would have to be done about the national scientific estate. Even if science–based initiatives were shown to be important in particular areas, it could reasonably be expected that such scientific activity would be supported by the individuals or firms who stood to benefit from it; in the case of health issues the same argument could be applied to charitable institutions.

It has been argued that only a dramatic demonstration of the potential significance of the contribution of science–based technical innovations could change the attitude of the liberal, 'laissez–innover' state (Salomon, 1973, p. 47). This demonstration was provided in the course of the Second World War. Although many facets of the war were deeply affected by the work of technical advisers ('scientific' methods were used in the development of codes, bombing strategies and convoy planning), it was the building of the atomic bomb which represented the greatest commitment of the state to the support of science. Here was a weapon which was predicated on principles unheard of until the years shortly preceding the war and which employed recently developed scientific concepts. It required the work of a great number of scientifically trained personnel in a large number of specialities and involved great expense – the UK was unable to afford the technical development at the time – which could only be found from the state's wealth. The development of the bomb was planned and carefully carried through. Here, to put it in a bizarre way, was a distinctively scientific, innovative product which only the intervention of the state's administrative and financial resources could bring into being. In Salomon's words (1973, p. 47), the bomb 'demonstrated that the time–lag between theoretical research and practical applications could be prodigiously shortened, if people were prepared to pay the price in men, money and logistics'. This instance strongly suggested that apparently arcane, basic research could have enormous practical implications and that the state might well have a strategic and economic interest in sponsoring such research.

Postwar economic planning took these lessons to heart (Freeman, 1974, p. 292), and a series of formal arguments were put forward justifying the extension of state support for a wide range of scientific activities in terms of their potential contribution to innovation. Of course the conditions of war were exceptional and special consideration had to be given to the issue of why state spending on science should be continued outside times of extraordinary need or emergency. The defence of spending on science was carried out in terms of a notion of market failure (Ronayne, 1984, pp. 38–40). The argument here was that, generally speaking, independent businesses would be inclined to carry out their own research and development if this would pay off in the long run with improved or wholly new products. To this extent technical innovation was just like any other business improvement such as managerial change or alterations in methods of financial planning. But, the argument runs, special features of science make this analogy inapplicable in large numbers of cases.

For one thing there are costs. An innovation may be profitable in the very long run, but firms which undertake the research are going to suffer *vis-à-vis* their competitors in the short term. If firms compete by trying to undercut each other then it will be precisely the ones which spend money on research which are likely to fail first. This is particularly true as scientific research becomes increasingly expensive. In some sectors firms may even be absolutely too small ever to make the investment in research facilities. The case of farms is often cited in this context; although they would benefit from systematic research on seed varieties, feed stuffs and so on, they have neither the budget nor the time to undertake these tasks. At the other end of the scale, some technologies require such enormous investments that no private firms could afford them. The research necessary for space flight and for the launching of satellites illustrates this problem.

The second problem stems from the difficulty of retaining exclusive use of technical knowledge once it has been gained. Economists refer to this difficulty as inappropriability. The vast expense of research may result in a product which can simply be copied; alternatively, staff with the newly developed knowledge can be poached. In such circumstances there may be great advantages in not being first to the market with a product. As we have seen from the examples supplied by Klass, in the case of the pharmaceutical industry companies rely on the protection afforded by patents to recoup the costs of research and development and, usually, to make a profit. But even patents are not foolproof.

109

They may be circumvented internationally or they may just be difficult to enforce.

The final reason why the market may fail to stimulate technical innovations is that investment in scientific research could be considered to be unusually risky. According to this line of thought, since success in scientific innovation is unpredictable, expenditure on research and development depends more on chance than other forms of investment. Consequently, companies are likely to undertake less than the optimum amount of research. For all these reasons, it is suggested, companies will tend to be conservative in their approach to research and development. The market mechanism alone will under–stimulate investment in scientific and technical research. Since the state has an interest in the wealth of its citizens and in the economic competitiveness of its industry the government has good reason to intervene.

State support for research may take a variety of forms. For example, agricultural research is often undertaken in governmental research institutions, and the positive results are passed on more or less freely to farmers. In other cases research expenditure may qualify for tax concessions which serve to decrease the effective costs of in–house research. There are also areas which governments traditionally present as in the national interest and in which they therefore have a prerogative to stimulate research. Foodstuffs are one case in point here. Another is military and strategic expenditure. Some forms of state-supported research have a double utility: geological mapping, particularly in the nineteenth century, went hand in hand with mineral exploration, and now deep sea oceanography can serve to identify locations in which submarines may be concealed.

Significant though these market failure arguments are, they have two weaknesses. In the first place they do not indicate how support for science should be allocated. The different facets of the market's 'failure' apply to different aspects of science and innovation. Thus, the concern over risk would be most applicable to basic scientific research which may have some practical value in the long run but whose utility is not evident at the outset. The first problem – the indivisibility of costs – applies to a wider range of research activities. To take the case of farmers, they may benefit from basic research on genetics but they are also likely to be assisted by scientific work of a more applied kind. Small farmers suffer from other indivisible costs as well; accordingly, they would benefit from assistance in sales and advertising in the same way as from assistance in research. The recognition of the likely forms of market failure provides general reasons for supporting science but does not specify closely how this support should be arranged.

Furthermore, the bases of the case for market failure have not escaped criticism. Mechanisms internal to the market may compensate for the weaknesses which supposedly lead to failure. In the case of inappropriability the mechanism of patents has already been mentioned. Ways can be found of combating the other problems too. Small companies can overcome the problem of costs by clubbing together to commission research from a university department or from private researchers and technical consultants. Equally, the huge costs of some projects can be met by companies forming consortia. Thus at present it is intended that the huge anticipated expenses of building a Channel tunnel from Britain to France will be met entirely from private finances. And where private money for such projects is not forthcoming it might be argued that this is not so much market failure as the market passing a negative judgement on a speculative investment. On this view, state money supplied to correct a supposed market failure may lead to people going ahead with a project for which there is no real call. Such a judgement has frequently been passed on a previous, notorious Anglo–French technical endeavour, the Concorde airliner. And this idea that state support may actually encourage projects for which there is no place in the market is used to handle the third aspect of market failure, risk. It can be argued that scientific uncertainty is no more extreme than other kinds of risk which firms have to face. And, anyway, why should it be supposed that governmental officials are any better at assessing a good risk than members of firms? Finally, it should be noted that the market failure arguments are premissed on a model of an economy composed of independent, competing firms. As we have seen a great deal of technically important production is now in the hands of giant corporations. These enterprises do possess their own laboratories and they are not in any straightforward sense susceptible to the dangers of market failure under perfect competition since they operate in oligopolistic or (sometimes) in monopolistic markets. They are not under the same pressures to be technically conservative although, as Gorz and Galbraith suggest, they will tend to address their technical resources to innovations which generate profits first and foremost rather than responding to consumers' requirements. For such companies the problem is not market failure so much as their success in the market.

111

National Science and Technology Policies: an Example

The arguments of economists suggest that there is a role for state spending to enhance a country's economically significant research. Analyses of market failure seem to apply well to instances where technical innovations stem from basic science which private companies may be unwilling to perform themselves. However, it is unclear to what extent states should support technical projects for which there is little enthusiasm in the market. As we saw in Chapter 3, national governments also have to meet the costs of basic research for which there are only, in Weinberg's terms, social and scientific justifications. What do governments do about the distribution of their support for science and technology?

One of the things they have done is to heed the advice of officials of the Organization for Economic Co–operation and Development (OECD). In the post–Second World War period the OECD has been steadily collecting information on national science and technology policies. It has carried out reviews of the science policies of particular countries which have been extremely influential in setting the pattern for subsequent national developments. One illustrative case is provided by the Irish Republic. After the war the Irish Republic, because of its neutrality and its escape from the destruction widespread elsewhere, did not receive the same aid with economic reconstruction as many other European countries. It had entrenched, protected industries and in most respects a conservative culture. In this period the government was also concerned to restrain state expenditure. There was thus little research and development spending either by the goverment or by the sheltered industries. When the OECD came to review Irish research performance (Department of Industry and Commerce, 1966) it was very critical of the limited science and technology expenditure. Even at the time of the follow-up review of 1974 the analysts (OECD, 1974, p. 34) considered it noteworthy that

> 45% of economically oriented R and D is located in a sector – agriculture – which accounts for about 17% of GDP, whereas the industrial sector – normally regarded as the science–intensive part of a modern economy – accounts for about 35% of GDP and only 42% of economically oriented R and D expenditure.

The recommendations of the OECD reviewers and the group of domestic policy analysts were subsequently largely implemented in establishing the science and technology policy apparatus.

It is important to note that the principal basis for the OECD reviewers' judgements was the experience of other countries. The policy proposals are defended by appeal to other countries' practices rather than by reference to points of principle. Thus, in the 1981 report of the Irish science policy body, the National Board for Science and Technology (NBST), a graph is provided showing how the proportion of national spending on research and development (GNERD) as a proportion of national wealth (GDP) compares with that of other OECD countries (NBST, 1982, p. 6). The stated aim of the NBST is to encourage Ireland to 'close the gap between Irish and European Gross Expenditure on R&D by the end of the decade' (NBST, 1982, p. 6). For most of the wealthy Western nations this figure is around 2 per cent. There is of course no proof that this is the right figure. But since the OECD proceeds by making suggestions on the basis of past reports, and governments are anxious to remain competitive, such a figure assumes a certain authority.

Among the statistics collected there also appears information on the bodies supporting and carrying out research. Information of this sort was provided in Table 3.2 for the case of the United Kingdom. There we saw that in 1983 the government funded virtually half of all research and development, while industry funded rather less but actually undertook nearly two–thirds of the national research effort. Again the figure of 50 per cent as a limit for governmental expenditure has taken on a symbolic quality; Ronayne remarks (1984, p. 73) that if it goes above this level 'this is usually regarded as a matter for concern, as the private sector should not be overly dependent upon government or overseas R & D'. However, as with all uses of statistical information in the social sciences, there are problems of interpretation here. The GNERD figure attributable to industry inevitably depends greatly on how companies define research. And as Klass reports (1975, p. 23), 'research' has special qualities:

> To the accountant, the word, since it spells magic for government tax returns, means any expense that can be categorized as a research expense likely to be accepted as a deductible item against taxation. This opens a vast amount of peripheral activity – as distinct from hard–core research in a laboratory. It may mean the costs of holding numerous conferences with travel expenses, not only for the research workers, but for surveys of research activities being done in competitive centres. It includes the establishment of large reserves of capital

to finance future research activities of the company. It may include the supply of very large amounts of expensive samples of the projected drug, during an early stage of its development, for tests on animals and for early trials on humans.

On this view, research covers anything from elaborate perks to capital acquisition and the internal purchase of the company's products on top of 'hard core' laboratory work. (This sceptical view of 'research' provides a further reason for treating published figures on research expenditure with caution.) The problem of pinning down what exactly research is constitutes a particular difficulty for the Irish Republic. In the last twenty–five years the Irish government has generally had a policy of expanding the industrial base by inviting investment from large foreign–owned companies. At the same time, successive governments have been eager to ensure that these companies carry on research and development activities in Ireland. The science budget statement for 1983, published by the NBST (1983, p. 6), welcomes foreign investment provided the companies 'embody key competitive functions such as R&D and marketing'. The report however recognizes that 'multinational companies tend to keep most high–skill operations in their home countries'. In order to satisfy the apparent demands of host governments these enterprises will inevitably be tempted to increase the amount of statistically classifiable research they do without necessarily carrying out their important research overseas. This issue has been addressed here because it highlights problems of science and technology policy formation; more attention will be focused on multinational companies' investments in Chapter 6.

Science and Innovation: Does Technology Derive from Science?

Although we have looked closely at the importance of innovations based on science and technology, little attention has been paid to the precise kinds of knowledge from which innovations stem. In so far as this issue has been mentioned at all the impression has been given that innovations are often based on new scientific ideas. This impression was reinforced by the case of the atomic bomb; a hugely powerful technology was generated from new claims about the properties of matter advanced by physicists. The significance of scientific innovations was also implicitly recognized in the theory of market failure. According to this theory the state should encourage or perform the scientific research which, in the end, will be crucial to future technical success. Equally, in Chapter 3 we encountered Weinberg's idea that scientific research can frequently be expected to have 'technological merit'. Underlying all of these ideas is the notion that technical innovations result from the application of new scientific insights and ideas. Such a notion is often referred to as the linear model of innovation (Ronayne, 1984, p. 44). The model is described as linear since it presupposes that first scientists make discoveries, then technologists exploit the new knowledge, and finally manufacturers develop new products. However, the pivotal role of scientific research in the generation of innovations had not, until quite recently, been systematically studied.

The eventual technical utility of science is a resource commonly employed by scientists to defend their apparently esoteric and obscure work (Potter and Mulkay, 1982). Instances like the atom bomb can always be pointed at to show that science does pay off in the end. But against this image of the utility of science it has already been mentioned that Tyndall was having to resist the charge, mounted by engineers, that science led people away from practical, useful skills. Moreover, as we have seen, the great mechanical achievements of the Industrial Revolution were far from dependent on the scientific developments of the eighteenth century. It has therefore been possible for various policy analysts and economists to challenge the idea that science is the 'mother' of invention.

In an early study of this issue, Williams (1964) used international statistical comparisons to suggest that, although there was a general connection between national wealth and the amount spent on research and development, the countries whose economies were growing fastest were not the ones spending the largest proportions on research and development. He stated (1964, p. 63) that 'there is no sign of a high correlation between *rates of growth* in output per head and the percentage of GNP devoted to research and development' (italics added). One possible interpretation of this finding is that science is a way of consuming wealth, not of amassing it. National wealth provides the opportunity to engage in the luxury of scientific spending but research and development do not necessarily benefit the economy. Such an investigation served to throw some general doubt on the contribution of science to economic success but hardly revealed the details of the connection between scientific spending and innovation. Subsequent studies attacked the linear model in two fundamental ways. The first line of criticism was that significant innovations did not stem from scientific developments so much as from technical innovations. Many innovations originated not at the bottom but a good way along the chain. The second assertion was that innovations were not driven by scientific developments; rather the new knowledge (whether scientific or more narrowly technical) arose in response to demand in the market or to a perceived need.

A whole series of studies have now been published investigating these issues; the series has been helpfully reviewed by Mowery and Rosenberg (1982) and Ronayne (1984, pp. 50–61). As Mowery and Rosenberg comment (1982, p. 193), these studies have 'been rather inconclusive – with results turning to a distressing degree upon the nature of the assumptions' of the authors. Two studies from the United States, probably the best known of all the analyses of innovation, serve to highlight many of the reasons for this inconclusiveness.

The first of these, 'Project HINDSIGHT' (Sherwin and Isenson, 1967), was carried out with the Department of Defense. By the 1960s a great deal of US science was being sponsored through this department under a compromise system which had arisen after the Second World War. The department was concerned to establish whether its investment in science and technology was paying off. An attempt was made to assess the contribution of research to the development of twenty weapons systems. The investigators identified 686 innovative ideas which had been important in the development of these systems. Fewer than 10 per cent of the events were deemed to have been scientific (as opposed to generally technical), and of those few only a minute fraction were said to have been due to undirected scientific research (Mowery and Rosenberg, 1982, p. 208). The apparent conclusions of this study were that technical innovations were far more significant than scientific ones, and that, of the scientific innovations, the overwhelming majority of useful ideas had come from applied scientific research. Basic scientific research appeared only very infrequently to be at the root of technical innovations.

The implications for the scientific community, and particularly basic researchers, appeared catastrophic. In response to the interim report of HINDSIGHT, the US National Science Foundation (a body overseeing non–defence and non–medical governmental research moneys) commissioned a parallel study of the crucial events underlying five twentieth–century civil innovations including the videotape recorder and the contraceptive pill (Ronayne, 1984, p. 56). The project, TRACES (Illinois Institute of Technology, 1968), was aimed at identifying the sources for these innovations in the fifty years before their production. In this case around 70 per cent of the key events were classified as due to basic research, and the majority of these were ascribable to research performed in universities and colleges.

Apart from the clear observation that the results of the studies were in conformity with the likely interests of the sponsoring agencies, the most striking thing about these studies is the extent to which the results depend on the starting assumptions and methods adopted. Where TRACES used a fifty–year cut-off point, HINDSIGHT searched back only twenty years; where the TRACES analysts selected clear product innovations, the HINDSIGHT team looked at the general improvement of weaponry; and where TRACES focused on products launched in a market, HINDSIGHT studied weapons built to specifications laid out in advance by the Defense Department.

In some respects the limitations which attach to these studies are reasonable; the Department of Defense could claim to be interested in only the sort of innovations which lead to weapons developments. But there has been a temptation to treat these studies as though they offered an overall answer to the question of science's role in innovation. They clearly do not. Even the apparent lesson for the narrow defence interest offered by HINDSIGHT is hard to interpret practically. Although it may be the case that the commonest source of innovation is technical research, if everyone adopted a strategy of supporting only technological studies the important 10 per cent of science–related innovations would be missed. Encouraging everyone to pursue the likely immediate advantage simply serves to reproduce 'market failure' kinds of phenomena. This line of reasoning is supported by the admission of the HINDSIGHT investigators that those science-based innovations which were of importance to the development of weapons systems tended to be of relatively great significance (Ronayne, 1984, p. 66).

One possible reaction to these studies is to try to formulate conclusions which make sense of both sorts of findings. For example, the results could indicate that there is usually a long delay before a new technical product can be devised which incorporates innovative science. Equally, it could plausibly be suggested that the majority of innovations are of a small scale and that these derive from technical changes; only occasionally do larger technical innovations arise and these are more likely to be associated with developments at a scientific level. But such conclusions can be expressed only in rather general descriptive terms; accordingly, and

117

as Gibbons and Johnston note (1974, p. 241), 'the very complexity of the relationship precludes simple calculations of the optimum size or distribution of the science budget'.

Some advance on these weak assertions can be made by looking at the work of Gibbons and Johnston. These authors studied the use of science and technology in the production of thirty innovations including a new type of quartz–halogen lamp, photo–sensitive inks and a non–dairy creamer (1974, p. 224). Their study distinguished itself from HINDSIGHT and TRACES by looking at the variety of sorts of information which were used in each innovation rather than trying to assign innovations either to science or to technology. The authors were then able to ask how much of the information coming from outside the firm responsible for the innovations was of a scientific nature; the answer was over a third. Although it was more difficult to classify the character of other information sources used by the firms, they conclude that of 'all information ... approximately one fifth' could be classified as scientific (1974, p. 230). Gibbons and Johnston also state that of the ten major external sources of information 'scientific literature had the greatest impact on the resolution of technical problems' (1974, p. 233).

This approach to analysing the impact of science on innovation frees us from the idea that science is important only if it is the direct source of the innovation. Gibbons and Johnston found that scientific publications were an extremely significant resource used in generating innovative products. They also found that contact with scientists, often with scientists working in the higher education sector, was an important facility in two out of every five cases (1974, p. 231). Such personal contact was responsible for the provision of a wide range of types of information. The importance of such contacts has recently been highlighted in a study of attempts at science and technology forecasting. It was considered that these exercises yielded a benefit more important than the actual forecasts: the contact promoted between 'academics, industrial researchers [and] policy–makers' (Irvine and Martin, 1984b, 144). Gibbons and Johnston additionally provide evidence indicating that scientists in higher education play an important long–term role in technical change since they will have determined the content of the training which the industrial scientists received.

In the present state of inquiry it is clearly impossible to produce a definitive general answer to the question, 'Does technology derive from science?' However, it is possible to summarize what is known about the relationship. First, scientists are routinely able to point to instances where a scientific development has apparently been vital to technical innovation. Such claims are a valuable defensive resource for scientists. To this end they often cite instances in which the scientific research was of a particularly undirected, basic nature (Yearley, 1984c). However, counter–instances can also readily be brought forward. But both these views are locked into affirming or denying the linear model of innovation.

Scientific research influences technical change significantly in ways which escape this linear model. Science is important in education, in the provision of informal advice and skills, for the publication of scientific information and also – although probably less often than scientists would wish to suggest – as a source of spin–offs when technologists are commissioned to build equipment for scientists. This last point was discussed in Chapter 3; there we saw how supporters of CERN were keen to draw attention to the economic benefits ensuing from the engineering contracts it issued. Indeed, two historians from CERN have recently argued that the 'socio–economic function' of the research centre 'cannot be systematically relegated to secondary importance' behind the issue of scientific merit (Krige and Pestre, 1985, p. 528). This function includes the 'technological fall–out' and 'long–term indirect effects on military technologies' which can be expected. On this view even apparently basic research is not without anticipated applications. However this assertion – like so much of the debate about the technological benefit of science – fundamentally depends on how terms are defined. Such decisions are inevitably contentious. Science, as practised in higher education and research establishments, undoubtedly does interact strongly with the economy, but there is, as Mowery and Rosenberg (1982, p. 216) note, a 'complex, nonlinear relationship'. A large element of this 'complexity' is due to the fact that assessments of technical utility and technological benefit are themselves socially variable. Both Gorz and Galbraith gave us reason to believe that technical merit is not something which can easily be gauged. To address this issue, therefore, we need to turn to the sociology of technology.

Note: Chapter 4

1   It should be borne in mind that Freeman expresses caution about the quality of the statistical evidence here (1974, pp. 199–200). The figures should be seen as indicative rather than definitive.

[5]

The Sociology of Technical Change: Lessons from Military Technology

Introduction

In this chapter the nature of technical change and the meaning of technical improvement will be examined. A discussion of technical advance at this stage will be helpful for two sorts of reasons. First, it permits the discussion of technical change in one very important substantive area, military technology. Second, it allows a comparison to be made between the emphasis of the earlier chapters of this book and the subsequent studies of innovation and the economic implications of science. Chapters 1 and 2 were concerned with the nature of scientific knowledge and the kind of authority which it could claim. The argument presented there was that in significant respects the authority of scientific knowledge is socially constructed. Decisions about the superiority of one scientific interpretation over another are not straightforward and uncontentious. On the other hand one of the main features of the last two chapters has been the emphasis on the practical utility of scientific knowledge. There may be room for doubt about the frequency with which economically significant innovations arise from scientific research, but the claim that scientific knowledge is at least sometimes economically beneficial seems to conflict with the idea that scientific knowledge is socially constructed. Some critical voices were raised in Chapter 4, including those of Gorz and Klass, but even they did not advance the idea that scientific and technical authority is socially constructed. Gorz pointed out that, despite manufacturers' claims, innovations may not be getting better in the interests of customers. None the less they are better from the producers' point of view. If, as his criticisms imply, betterness can be assessed straightforwardly, it is hard to see how technical decisions could be said to have a social component. Furthermore, a non–constructionist view of technology would have negative implications for the sociology of science also, since, as described in Chapter 3, science and technology are becoming less distinct as jobs, practices and forms of knowledge. For reasons both of principle and of substantive interest we therefore need to study the sociology of technical change.

Technology and Society

The customary and in many respects most obvious way of thinking about the relationship between technical change and society involves looking at the impact of technology on society. A typical focus for interest here would be a concern with the alterations produced in social mores or social relations by technological change operating as an external factor. Clear and familiar examples could be drawn from the social impact of medical technologies like vaccines and contraceptives, of transport technologies, or of productive technologies. If the reverse relationship (that is, society's effect on technical change) is considered at all within such an approach it is usually in terms of the facilitation of the technical change by the receiving society; there might, for example, be interest in the social basis of resistance to the introduction of vaccines or to water fluoridation. The view that technology has the dominant role in its relationship with society is even echoed in economic studies of technical change; market failure theories, for example, are concerned about the social provision for research but do not look at the impact of society on the content of the innovation.

More recently analysts have attempted to focus attention on the other aspect of the relationship: the influence of social factors on technical choice. The principal way in which the status of technology as an external influence has been undermined has been through studies of the political economy of technical decisions. The argument underlying such studies is a familiar one. In essence, it is that one technical artefact, design, or procedure has been chosen over another not so much or not only for technical reasons but also for economic or political ones.

121

It will be helpful to consider two clear examples of this form of analysis. The first, which has already been mentioned in Chapter 4, concerns the reasons behind mechanization in the agricultural implement manufacturing industry. According to this study, pneumatic moulding machines were introduced to McCormick's foundry in Chicago, where reaping machines were produced, towards the end of the last century directly on the initiative of the proprietor. The new machines mechanized the production process and might therefore be thought to have been introduced for reasons of cost saving, greater accuracy and reliability. Yet 'the new machines, manned by unskilled labor, actually produced inferior castings at a higher cost than the earlier process. After three years of use the machines were, in fact, abandoned' (Winner, 1985, p. 29). The analyst's claim here is not that technical change promoted deskilling *en route* to efficiency but that new machines were brought in just to break the strong power position commanded by the skilled workers of the National Union of Iron Molders. In the original study on which Winner draws, Ozanne argues that the company's interest in the machines definitely centred on their ability to displace labour. In support of this claim he cites the fact that the company uncharacteristically bought a complete set of untested machines rather than acquire experimental ones first (1967, pp. 26–7). It is the technical *inutility* of the innovation which supplies the evidence that the change was not technically motivated. In this instance technical change seems to have been directly determined by social and political values.

The second example is the well-known study by Ruth Schwartz Cowan (1985) of the development of refrigerating machines which she relates as a just–so story about how the fridge got its hum. She points out that before such machines became widely spread in the domestic market there were two rival design principles. In both cases the task of removing heat, from the cooling compartment was achieved by continually vaporizing and then recondensing a liquid. The first process absorbs heat, while the second releases it. In this way liquids and gases can act as the medium for transferring heat from one place to another. The difference lay in the energy source which was used to drive the cycle of condensation. In one case the machinery was gas powered while in the other, as in the vast majority of modern machines, it was electrically driven.

The very earliest machines on which the principles of heat transfer were worked out were huge devices mostly used in commercial locations. They tended to be designed or at least modified for each particular installation. And as the power source was often contained in a different housing from the cooled container it was easy to appreciate that the choice of the form of energy to drive the machine was open; indeed, early machines used in breweries or in meat storehouses were often steam powered. Subsequently, the attempt was made to provide refrigeration units for domestic applications. This required a much more compact power source. At the same time it provided the basis for a very large market for a standardized set of products. Competition increased to have one particular system adopted as the standard. In the cases of both gas and electrical appliances technical innovations were required. The task of supplying an integrated domestic unit was a challenge to both sets of designers. However, there were additional parties whose interests were at stake in this matter: the public utilities companies. The choice between widespread adoption of gas or of electrical refrigerators would make a large difference to the demand for the products of the privately owned United States power supply companies. Thus the choice between the two designs – the deciding factor in the struggle to determine which design became most successfully domesticated – was not their technical performance but the aggressive support of one design by the electrical supply company. Cowan actually claims that in many respects the gas machine was superior for domestic use; it was quieter and, crucially, had fewer moving parts to go wrong.

The political economic critique of technical choices thus draws its force from being able to demonstrate that the machine which was actually inferior was promoted over the superior one for extrinsic reasons. The gas fridge was preferable but was squeezed out by shrewd market manoeuvres; the new foundry installations were technically inferior to the preceding process but had desirable short–term political consequences from the employer's viewpoint. The curious thing about this approach is that it tacitly acknowledges the premiss of the model of technical change which it sets out to criticize: that there is one single best technology. By showing that *in this case* there was a disparity between the technically preferable design and the product actually chosen, this approach operates to reserve a special status for the issue of technical choice. The claim is that extrinsic considerations overrode properly technical ones. It is thus implicitly argued that, in the absence of distorting circumstances, decisions about technology would be uncontentious. This position makes technical choice *asocial* and uses evidence for the social contamination of choice as the basis for criticizing the selection of particular technologies.

Analytically the situation here precisely mirrors that of the sociology of science outlined in Chapter 1. As was noted then, some sociologists have concentrated on external social influences on scientific development, on occasions when the theological, social, or political implications of scientific beliefs have led scientists to override scientific considerations (as was apparently the case with Lyell's decision about the evolutionary history of mankind). Such an example is analytically identical to McCormick's preference for trades–union–busting technologies. In both cases it is implicitly conceded that proper scientific or technical choice is, or at least could be, asocial; these instances are the exceptions which prove the rule. While such cases may be of great practical importance they are not analytically fundamental since they do not question the principle of unambiguous scientific or technical choice. In Chapter 1, Shapin's case study of the science of phrenology and Collins's analysis of physics were used to argue that even internal scientific choice has a social dimension. A fundamental reappraisal of the technology and society issue requires an 'internal' sociology of technical decision-making. A case is needed not where social or political considerations override the technical but where they are infused into it: where technical improvement is the objective but where the nature of 'improvement' is disputed.

Technical Improvement in Military Technology

The example of one area of technical development – defence or weapons technology – will allow us to examine the disputes over improvement in considerable detail. We have already seen that in the Western world very many scientists are employed in defence work and that, in recent years, over a half of the UK government's research and development expenditure has been devoted to this area. Defence work is undertaken by scientists in government research facilities and in private industry. The amount of scientific effort which is expended on defence varies considerably from country to country. Military acquisition by some nations like Japan and the Federal Republic of Germany is limited by constitutional means or by international agreement. Other aligned countries have vast defence budgets of which a considerable proportion is devoted to research and development. In 1981, for example, the USA devoted 5.8 per cent of GDP to defence, while the UK committed 4.9 per cent and France 4.2 per cent (Gummett, 1984, p. 57). Even neutral countries are not necessarily spared such expenses. While some, like the Irish Republic, spend only small amounts on defence and devote most of this expenditure to personnel, others choose to defend their neutrality by arming heavily – as is the case with Sweden, which supports a large armaments industry (Howe, 1981, p. 335; Smith and Smith, 1983, p. 73). In general, however, the world's major economies – countries which are wealthy in absolute terms rather than simply wealthy by head of population – tend to be large producers and traders in armaments. Worldwide, therefore, an enormous amount of research effort goes into military technology.

To a certain extent the amount of expenditure on weapons systems can be explained in straightforward terms. Countries which wish to retain their political sovereignty are obliged to have armed forces to defend their interest if necessary. Since armaments by definition are for use in competitive circumstances there is a constant pressure to improve weaponry in order to match the level likely to be attained by one's potential adversaries. Governments, as purchasers of weapons, and the producers of armaments thus have a common interest in generating innovative weaponry which will keep their armed forces ahead of the competition. Given this common objective and given that the purpose of weapons is rather narrowly defined, one might expect that technical improvement would be easily recognized and criteria for improvement readily agreed. Evidently, no one benefits from making worse armaments, so the presence of social elements in decisions about weapons improvement might seem extremely unlikely. Yet there is room for dispute over what should count as an improvement. The production of improved weapons turns out to be of considerable sociological interest.

The obligation to keep weapons systems ahead of competitors leads to a constant demand for improvement. To some extent this improvement can take the form of increases in numbers; commanders will generally welcome extra tanks or more ammunition. But improvement can also, and frequently does, take the form of product innovations: new aircraft, missiles, guns, or even the provision of a defensive shield as envisaged in the US 'strategic defence initiative' (Kaldor, 1983, p. 130). As is the situation with other research, like the basic research described in Chapter 3, the demands of improvement are costly, and the price of military research rises and rises ahead of inflation. The production of state-of-the-art military equipment is extremely research intensive; according to a Ministry of Defence review cited by Gummett (1984, p. 73), 'production/R&D ratio[s] of around 3:1 or 4:1' are

125

commonly estimated in the defence industry although in practice these ratios have tended to be even smaller. A very considerable part of the cost of innovative weaponry thus lies in the research and development expenditure. Such expenditure alone accounted for over two–thirds of one per cent of the UK's GDP in 1981 (Gummett, 1986, p. 61). The new products are also extremely expensive in themselves; as Gummett (1984, p. 75) states, 'the cost of projects, whether wholly British or collaborative, has risen inexorably to such an extent that it is doubtful whether any European country alone can afford the next generation of major aircraft'. In a similar vein Kaldor (1983, p. 18) cites a droll estimate that on current rates of cost escalation the US Air Force's complete budget will buy only one plane by the year 2020.

Noting the vast scale of weapons expenditure in virtually all Western nations, sociologists have sought to investigate whether there is some additional social function performed by this massive spending. One such interpretation has famously been put forward by Baran and Sweezy (1968, pp. 178–214). These authors propose that the role of military spending and military might should be examined in the light of the interests of capitalist nation states. In their view, and contrary to proclamations about international co–operation, capitalist nations are inherently in a state of rivalry and potential conflict with each other (1968, p. 178). Thus, in the two centuries preceding the British military supremacy of the Victorian period, warfare was the constant accompaniment to attempts at expanding empires and the pursuit of trading and slave wealth. In this context international peace is regarded as an extraordinary condition which arises only when one power has attained undisputed dominance and when it can manage incipient conflicts elsewhere. The nineteenth–century British peace gave way to a prolonged period of international conflict which was finally stabilized after the Second World War by United States predominance (1968, pp. 180–2).

Baran and Sweezy interpret contemporary military expenditure as important in a number of ways. First, it is a form of force used by capitalist (and, one should add, state socialist) countries to support their economic, strategic and political interests around the globe. These authors contend that the US capitalist system is intent on destroying Soviet–led socialism. In offering this interpretation they are led into doubtful and conspiratorial assertions about the intentions of 'the American oligarchy' (1968, p. 190). It is impossible to lend any credence to the idea that there is a single, unified American oligarchy. It is, however, entirely reasonable to maintain that high levels of arms expenditure in the USA are repeatedly justified in terms of dire but often questionable and unsubstantiated warnings about the Soviet threat (Smith and Smith, 1983, pp. 20 and 49). More particularly, military support is offered to friendly or, at least, anti–communist governments, often in response to a perceived danger from revolutionary socialist movements. Armed forces are also employed directly to police international trading, as the presence (in the summer of 1987) of French, US, British and Soviet warships in the Iranian Gulf indicates (*The Guardian*, 23 June 1987, p. 8). Finally, they suggest that military expenditure fulfils a domestic economic function. They claim that military spending, a legitimate government monopoly, is used as a device for managing the economy. Such an opportunity for expenditure operates as a form of economic safety valve which can be used to cope with the booms and slumps, the crisis tendencies, of the capitalist economy.

On this view the state can, for example, promote economic activity at times of economic slump through the commissioning of new military hardware. As Baran and Sweezy state (1968, p. 211): 'Here at last monopoly capitalism had seemingly found the answer to the "on what" question: On what could the government spend enough to keep the system from sinking into the mire of stagnation? On arms, more arms, and ever more arms.' Certainly naval orders have regularly been the salvation of shipyards, and contracts for the manufacture of tanks have helped US automobile companies facing difficulties (Kaldor, 1983, p. 134). For Baran and Sweezy the state can thus assist and regulate vital areas of manufacturing activity without intervening directly in the general market system. One might also add that the military is not without direct influence on the employment market. The idea of 'spin–offs' or 'trickle down' from military technology can also be deployed to suggest that military expenditure is a means for the state to ensure that new technologies are available to industry. This is, it should be noted, a variation on the market failure theory. On a Marxist interpretation, such spending functions to counteract the consequences of the tendency of the rate of profit to decline.

Even for Baran and Sweezy, however, this solution to the 'on what' question appears imperfect. They cite two reasons for the dwindling effectiveness of the safety valve function of military spending (1968, pp. 211–14). They mention first the rising research and development commitment which was noted in the last section above. To Baran and Sweezy this suggests that the amount of investment available to be directed to the macroeconomically significant aspects of armaments (that is, those aspects which demand vast amounts of materials and labour, such as shipbuilding) will diminish. Secondly, they invoke the idea that the arms build–up is beginning to lead to a decrease in security (1968, p. 213); they suggest that 'responsible leaders of the United States oligarchy' perceive the futility of further arms development. Neither of these arguments seems particularly persuasive.

The attractions of the second claim seem to be undermined by recent developments, notably the 'strategic defence initiative' (SDI) or 'Star Wars' scheme favoured by President Reagan. The innovative and fantastically expensive idea of a defensive shield is justified in terms of the increase in security it supposedly offers (E. P. Thompson, 1985b, pp. 15–19). Whatever one thinks of the value of SDI it cannot be doubted that its advocates present it as an improvement over existing nuclear defence strategies. Influential elements of the US 'oligarchy' are clearly no longer thinking along the lines Baran and Sweezy imply.[1] The other argument can be answered as follows: because technical innovation is so central to economic success, military expenditure on research and development could just as easily be functional for the economy as militarily inspired expenditure on tanks, boats and other hardware (Galbraith, 1974, p. 233). It is by no means clear that such a shift in the nature of arms spending would make the military sector any less viable as a safety valve.

A final challenge to the argument about the economic benefits of the permanent arms economy can be mounted by questioning its functionality. One can note, for example, that the Western economies in which the highest percentage of national wealth has been devoted to military expenditure in the postwar period have generally experienced the slowest economic growth and greatest problems of unemployment. In proposing this argument Smith and Smith (1983, p. 88) show a generally inverse correlation between the commitment to military expenditure and economic success. Countries where high percentages of national wealth are devoted to military spending, the UK and USA, have the poorest economic record. Conversely those with the best postwar economic record, the Federal Republic of Germany and Japan, have comparatively low levels of spending. Such figures, of course, indicate only an association; they do not demonstrate that levels of military expenditure and economic success are causally related. None the less, Smith and Smith (1983, p. 87) argue that there is a deeper connection, since 'it is investment or fixed capital formation – the purchase of plant and equipment for use in production – which is the major component of demand to be displaced by military spending'.

Other commentators have proposed more specific dysfunctions arising from large commitments to military production: the concentration of skilled technologists in specialist military centres insulated from the civilian economy (Milne, 1984, p. 12) and the unsuitability of military research for incorporation in civil projects (Gummett, 1986, p. 62). We will briefly return to these issues at the end of this chapter. At this point it is sufficient to note that the case for the functionality of weapons spending is far from compelling (MacKenzie, 1983, pp. 51–3).

The Military–Industrial Complex

Rather than interpret continued high levels of military expenditure in the light of the supposed requirements of the capitalist state, some authors have looked to the structure of the arms manufacturing and purchasing system. The point of interest for these authors is the way in which military spending benefits particular companies and sectors of the government and armed forces. On this view, arms spending is not of value to the state as a whole; rather it is in the material interests of a group, the members of which act in a concerted manner and often lobby for their interests. There has long been evidence of the practical autonomy enjoyed by arms producers; Howe (1981, p. 317) relates that the British paid royalties to Krupp of Germany for explosives equipment used against Germany in the First World War. He records (p. 316) that arms merchants 'even sold weapons to their country's enemies to improve the chances of greater domestic sales'. Equally, a spirit of even–handedness is evident in the recently reported deal by 'Japanese companies to sell advanced machinery to Russia to help Soviet submarines escape detection by Western patrol craft' (*The Independent*, 24 June 1987, p. 8).

However, this issue came to prominent public concern in the USA in the 1960s not because of fears of outright military disloyalty but on account of more insidious dangers. It began to be felt that arms manufacturers and those who commissioned weapons often operated in league and in a way which was unaccountable to the public. The anxiety was that there was forming a military–industrial complex (MIC) making decisions about the spending of public money and the arming of the nation immune from public scrutiny. Such a view was voiced by President Eisenhower who warned of 'the acquisition of

unwarranted influence ... by the military–industrial complex';
the President, as Berghahn remarks, 'could be assumed to know
what he was talking about' (1981, p. 86).

The MIC, it was suggested, was made up of the principal
contracting firms which regularly supply the US forces, highly
placed service personnel and staff of the Department of Defense.
The danger presented by the complex is the possibility that,
since the firms regularly receive orders from the officials and
the officials are the chief customers of the firms, the members
of the MIC may be inclined to develop weapons in a way which
suits their personal and commercial interests rather than national
security. Their ability to behave in this way is enhanced by two
aspects of the weapons industry: its technical specialization and its
secrecy. To begin with the latter, it is clear that for reasons of
security, information about the design and choice of weapons has
to be safeguarded. But the same secrecy which protects national
interests can cover up waste, poor designs and inefficiency. Second,
since 'the business of inventing, acquiring and producing weapons
[has become] a scarcely comprehensible process' (Erickson, 1971,
p. 227), the technical authority of arms specialists has meant that
outsiders can have little say in matters of arms procurement. On
this view, because outsiders could not understand high–technology
defence they could not judge it.

Various signs of the dangers inherent in this situation started
to come to public attention. First, there were delays and related
inefficiencies. New weapons systems which were supposed to be
produced with all possible haste to meet the Soviet challenge ran
over the production deadline. Second, it was found that a huge
number of defence contracts were handled by very few companies.
It seemed that there was little practical competition for defence
contracts. In turn these few companies, the prime contractors
(Kaldor, 1983, p. 10), were often heavily dependent on continued
defence contracts for their very survival. Companies which carried
on over half their business in the area of governmental defence
procurement were 'locked in'; they were highly dependent on
gaining further orders. Ironically, this dependency also threatens
to compromise the government, since any decision which denied
further contracts to these companies would have huge economic
implications for the region in which the firm was based. For this
reason North American firms (and other interested parties) were
able to exert a great deal of pressure on their regional representatives
in Congress to ensure that orders continued to come their way.
Even unions have an interest in the maintenance of orders and

have lobbied alongside armaments companies (Howe, 1981, p. 354). In the view of other politicians (such as the Democrat T. J. Downey, quoted by Howe, 1981, p. 355), contracts may represent 'military–industrial welfare program[s]' aimed at avoiding unemployment and economic decline in particular regions.

Furthermore, in addition to the relative security of receiving continuing orders, the arms producers often enjoyed large financial rewards including high rates of profitability. According to a report in the *New York Times* from a Senate investigation in 1962 (Baran and Sweezy, 1968, p. 206), 'Boeing's profit ... was "almost double" the 10.73 per cent average net profit' for US manufacturing industry in the period preceding the inquiry. The economic benefits generally arose from several sources. Since contracts have to be awarded before new equipment is built, and since, by definition, the equipment is innovative, costs cannot be precisely predicted. Contracts could not therefore be for a fixed price. Often they were agreed on a 'costs-plus' basis whereby the company was guaranteed the costs involved in devising the new weapon plus an agreed percentage for profits. Under such an arrangement there is little incentive to limit costs; rather the opposite. Thus the tendency was for final costs to far exceed initial estimates (Pavitt and Worboys, 1977, p. 28). The fact that companies and the budgets of the purchasing agencies are, to a greater or lesser extent, protected by secrecy has meant there is less of a check on these price overruns than in other industrial sectors. Contractors can also increase their effective profits in various ways. Bearing in mind Klass's remarks (cited in the last chapter) about the manipulability of 'research' expenses, it is easy to envisage how funding for a defence contract can be used to subsidize other parts of a contractor's business (MacKenzie, 1983, p. 38). Third, because the equipment is for national defence there is a tendency for components of the highest conceivable standards to be sought. Even if these are truly the best and of practical benefit, such 'gold–plating' of the design inevitably forces up the costs. Under a costs–plus arrangement gold–plating clearly holds other commercial attractions. Finally, the uncertainties inherent in the contracting system can be exploited. Howe (1981, p. 342) cites the following example:

Contractors [may] buy into a contract by making a bid they know to be too low. Once work has started, it is not usually difficult to get the Pentagon to agree to reassess costs ... A classic example of 'buying–in' was Lockheed's contract to build the first C–5A transport [plane] – the world's largest – for $1.9 billion, undercutting Boeing and McDonnell–Douglas. The final cost was $3.9 billion.

Still More Complex

Attempts were made to document and combat the problems of spiralling costs, procurement delays and the pressures exerted by locked–in companies. One Senate investigation has already been mentioned; Proxmire (1970) relates many of the critical findings of the Subcommittee on Economy in Government of the Joint Economic Committee of Congress. By 1976 the US government had reformed the contracting system so that the specifications of weapons systems were more closely agreed in advance and had revised the simple costs–plus arrangement (Howe, 1981, p. 344). The mutual dependency of the locked–in relationship in the USA has also been reduced so that 'by 1975 the top 25 arms producers' military work accounted for less than 10 per cent of their total turnover (compared to 40 per cent in 1958)' (Smith and Smith, 1983, p. 72).

In these various ways the coherence and unity of the MIC appeared to be breaking down. None the less many of the central features of arms purchasing have persisted. For example, there still remains great uncertainty in the acquisition of new weapons, which can readily lead to cost overruns and delays. Even if firms are now penalized in some manner for such occurrences, the purchasing agency still has to meet the bill. Having argued that some item of defence equipment is necessary the government is hardly in a position to forgo it. This situation can be exemplified by the recent British example of the Nimrod airborne radar system. This British–built system was ordered in 1977 with the expectation that it would come into service for training use in 1982 and into full service in 1984. By December 1986 Nimrod was still not ready. It was then cancelled, in the words of the Secretary of State for Defence, 'notwithstanding the expenditure of £660 million so far, just over half of which has been spent on the avionics [aviation electronics]' (House of Commons, 1986b, p. 1350). A US system, based on Boeing aircraft and costing around £1 billion, was purchased in its place. Even in this case, where the government did not simply tolerate the delays associated with the home–grown technology, huge costs and some measure of delay were not avoided. The strict provisions for secrecy over defence matters in the UK prevented the technical difficulties, delays and escalating costs from becoming generally known sooner. Moreover, even in the parliamentary debate over the Nimrod decision, it was still being urged that no more costs–plus agreements should be made in the UK (House of Commons, 1986b, p. 1429).[2]

The case of Nimrod illustrates further influences on defence procurement decisions. The day before the final verdict on the planned adoption of Nimrod by the RAF was announced, a story was circulated that the system had been 'given secret trials by the French', who were also seeking an early warning system (*The Independent*, 17 December 1986, p. 1). One implication which could readily have been drawn at the time was that if the British government failed to purchase the native product then the French were hardly likely to. Equally, on the day of the negative decision GEC (the principal avionics contractor behind Nimrod) warned that 2,500 jobs were at risk in their own company and among subcontractors. Union estimates were said to put the figure closer to 3,500 (*The Independent*, 19 December 1986, p. 4). Such considerations cannot easily be overlooked, and it is hard to avoid the conclusion that they are mobilized precisely for that reason. Of course, this strategy is potentially available to both sides in any dispute. In this case it was made clear that the purchase of the aircraft from Boeing would result in Boeing subsequently offering contracts to British companies worth around £1 billion.

A third factor which continues to affect defence procurement relates to the issue of technical authority, mentioned earlier. Both because of their practical experience and because of the limitations of official secrets, the only people in a position to evaluate defence requirements and technical possibilities are employed in arms companies, departments of defence, or the armed services. When attempts are made to increase competitiveness in contracting it is important that these agencies maintain their independence and pursue their proper organizational interests. However, the actual relationship is frequently incestuous. Howe relates that in 1959 an inquiry revealed 'that 72 US defense contractors were then employing 1,426 ex–officers, including 251 of general or flag rank' (1981, p. 333). By the 1970s little had changed (1981, p. 352):

> All prime contractors were found to have on their staff at least some former Defense Department officials or officers whose previous tasks had been to evaluate the work of their present employers. Between 1974 and 1975, the number of former

133

senior Pentagon officials working in the defense industry increased from 433 to 715.

The opportunities for 'incestuous' contacts are manifold. Retired service personnel may enter the staffs of firms where their job will be to persuade former colleagues to purchase their new employer's products. Experts from the industry may be seconded to the defence department. Civil servants may be poached by arms industries. Additionally, elected representatives, like Members of Parliament in the UK, may be senior employees or directors of weapons manufacturers. This restricted élite necessarily has many of the properties of the core set of authoritative decision-makers described in Chapter 1. In this case, however, there is the constant potential for a conflict of interests to arise.

Finally, and often relating to conflicts of interest, there is the issue of deliberate overcharging and corruption. There was the recent case of a charge of $800 being made for a pair of pliers (*The Guardian*, 23 May 1985, p. 6). And a few days later the US Defense Secretary

announced that three US Navy officers, including an admiral with 33 years' service, are being relieved of their duties because a supply depot under their command paid the Grumman Aerospace Corporation $659 (£519) for each of seven aircraft ashtrays.

(*The Times*, 1 June 1985, p. 6)

In addition to such rather small efforts to inflate the budget, overcharging may arise in the context of gold–plating. Since a policy of 'only the best' can be justified for machinery of great technical excellence, unnecessarily expensive items may be used. A bizarre case, which nevertheless exemplifies this tendency, is related by Turner. Several hundred coffee machines were bought by the US Air Force for an unusually high price: 'Justifying the purchase, a Pentagon spokesman said: "The brewer ... is to be installed in the Lockheed C5A. It is a very reliable device and will continue making coffee after loss of cabin pressure following a direct hit"' (Turner and SIPRI, 1985, p. 61).

There is also the possibility that members of the technical élite will receive offers of inducements to make decisions in the favour of a particular company. An illustrative example is reportedly provided by the involvement of Admiral Rickover of the US Navy in the dealings of the General Dynamics Company.

This company was charged with '"pervasive" business misconduct' and, in May 1985, was fined over $600,000. It also had some contracts cancelled and suffered (*The Guardian*, 23 May 1985, p. 6) 'the suspension of two of its divisions from obtaining fresh contracts until it repaid $75 million in overcharges and instituted a new code of ethics for its staff'. *The Guardian* report also claimed that Admiral Rickover was publicly censured: 'Admiral Rickover's offence was that he had apparently demanded expensive gifts – including earrings for his wife and plastic laminated $50 bills – totalling $67,628 from General Dynamics.' This report claims, however, that despite the apparent seriousness of the misconduct and the large fine, General Dynamics will be able to 'restore its normal working relationship in a matter of weeks'.

It appears therefore that many of the principal conditions which favoured the development of the MIC have not been altered and that defence procurement is susceptible to the same sorts of pressures as encouraged anxiety in the 1960s. As Smith and Smith point out (1983, p. 74), these pressures arise not chiefly from conspiracy or bribery but from the mutual interests of weapons manufacturers and arms consumers in generating new and better weapons. It is to this concern with improvement and the meaning of 'better' that we should now turn.

The Social Construction of Technical Merit in Weapons Innovations

As has already been noted, a number of aspects of military technology might lead one to suppose that technical innovations would be readily agreed. In some sense the warlike objective of weapons systems is singularly clear. Moreover, we have just seen that there exists a common interest among the relevant parties in devising innovative weapons and that decisions about weapons procurement are made by a core set of people with technical authority. Can there be any room for social construction in the choice of weapons? Clearly, there is room for social influences in the form of malpractice and in the pursuit of one's firm's or department's interest. But such influences are extrinsic to the technical assessment of weapons. Bribes, one would suppose, are most necessary when the attempt is being made to persuade the government to acquire a product of doubtful quality. Can we talk of the social construction of technical merit; that is, can we talk of social variability in assessment of the intrinsic quality of weapons? A number of features of weapons systems mean that we can. Interestingly, these features turn out to be very similar to the conditions surrounding the evaluation of modern scientific results (about neutrinos and gravity waves) discussed in Chapter 1. Just as with experimental science, we need to go into some detail in order to discern the variability. Many of the cases employed in the following discussion are drawn from Kaldor's vivid account (1983) of the details of contemporary military technology.

The first feature of military technology which allows us to talk about the social construction of merit is the difficulty of working out a suitable test for such technology. This is analogous to scientists' difficulties in devising appropriate experimental equipment for detecting wayward particles. The test is only an instrument for picking out the desired qualities, and a judgement about those qualities is inevitably built into the test. Thus, while it may be relatively easy to accept that a plane has been improved if it now goes faster, speed can mean various things. As Kaldor points out (pp. 132 and 136), while the F–15 fighter flies extremely quickly, the mean 'time between failures [is] eighteen minutes'. Since so much of its time is spent in servicing it is impossible to determine a single true speed against, say, a slower but more reliable aircraft. A similar point is made by Fallows (1985, p. 241)[3] where he reveals that tests of an automatic rifle were carried out by:

> sharpshooters and marksmen who measured a weapon by how well it helped them hit a target ... six hundred yards away [and by technicians who claimed that] if it couldn't be fired in the Arctic and the Sahara and perform just as well in each place, it was not fit for army duty.

Tests can be set up in a great variety of ways and the results of technical tests, just like the outcomes of experiments, are subject to interpretation.

One way of overcoming this open–endedness is for the measurement of performance to become institutionalized along certain dimensions, often ones which are readily quantified. Aircraft speed has already been mentioned. Kaldor (1983, p. 19) gives the related example of the Trident submarine which boasts a higher speed than its predecessors but whose 'top underwater speed (25 knots) is still significantly lower than that of attack submarines (30 knots), and [which] in any case ... is so noisy at top speed that it would have to go slowly to escape detection'. A similar phenomenon can be observed for the case of missiles, whose accuracy of strike can be measured using statistical mathematics; an assessment in these terms precludes other considerations (such as the hardiness or reliability of the guidance system) affecting the likely utility of the missiles (p. 139). Equally, another quantifiable dimension along which aircraft can be evaluated is the amount of weaponry they are able to carry. However, the amount they can in principle carry may not be at all closely related to the amount that the pilot can manage to use in combat (p. 137). And, in time of war, the combat effectiveness of air assaults may even be 'measured' in terms of the *number* of bombs dropped (p. 157) rather than in relation to the targets hit. Whilst, in these cases, the value measured has some plausible connection to technical merit, the choice of this measure as a representation of merit is clearly just one among many ways of closing down the openness of technical assessments.

Furthermore, even if these dimensions of measurement are accepted as good indices of merit, the measures themselves depend on the same sorts of negotiation as were described in relation to scientists' experiments in Chapter 1. If a missile or a plane (like the F–15 mentioned above) is subject to malfunctions, are these to be calculated into the measurements of speed? Does one average only from 'successful' flights or from all flights? Human judgement is required. A further, related issue concerns the transition from test conditions to actual use. As well as being unreliable, innovative weapons tend to be more complex. This means that they will often be too complicated to repair under the conditions of combat; worse still, more complex weapons make greater logistical demands on the fighting forces. Troops need to carry greater numbers of spare parts and they require better supply routes (Kaldor, 1983, pp. 131 and 135). The range of support services required by a new weapon is a factor which is hard to build into readily measured tests of merit; it is also practically impossible to predict the detailed importance of this factor in advance of actual combat conditions.

A fourth feature of modern weapons systems which enhances the effect of the issues so far mentioned arises from problems encountered in testing. There is an understandable reluctance to waste new, expensive weapons on lengthy tests and even on the training of novices. As Kaldor notes (p. 136), 'Many soldiers, sailors, and airmen are familiar with the problem of being allowed only one live firing of their principal weapon per year.' Such economies are easy to understand if one notes that the air–to–air missiles which have replaced machine–guns on aircraft have increased the cost of trial firings by 'a factor of tens of thousands' (p. 154). At the most awesome level it is clearly impossible to test nuclear missiles under 'realistic' conditions. Indeed, one of the disputes of recent years about the siting arrangements for US long–range nuclear missiles has been fuelled by the impossibility of test. Some precaution has to be taken to protect missiles against a pre–emptive strike. One proposed solution is to locate them very close together (a proposal known as 'dense packing': E. P. Thompson, 1985b, p. 22) in the expectation that the disturbance caused by the first incoming missiles would disrupt the final approach of subsequent ones, thus effectively shielding some of the densely packed US weapons. Such a defensive strategy was, naturally, quite beyond testing; the effect of a huge nuclear explosion on subsequent incoming missiles is just not known. In this way the practicalities of using new weapons may remain unknown and their war–readiness quite unproven. Without even the room for tests there is bound to be a great deal of room for legitimate differences of opinion about the merit of new armaments.

Kaldor describes military technology as increasingly 'baroque' (1983, p. 2), meaning that it is more and more elaborate and almost ornate. This element of over–elaboration is partly composed of the factors so far discussed. But two further features are particularly significant in this respect. The first is the possibility that attempts at improving all aspects of a technology, of gold–plating the whole design, will lead to a product of exaggerated complexity. Kaldor cites (p. 137) an innovative aircraft which has so many features that the pilot cannot use them all. Quoting from a report in the *International Herald Tribune*, she notes that the F–15 'pilot "flies the plane and directs its two missiles, one for targets out of sight, another for those within sight, and fires its Gatling gun at targets nearby" and, presumably, he also has to maintain communications, etc'. It may be that in every dimension considered individually the plane could be argued to be better than its predecessors. But if, in total, it baffles the pilot – particularly under the pressures of combat – it is difficult to see how it truly is technically better.

If the gold–plating tendency leads to bizarre elaboration by one route, the international and inter–service demands on weapons production generate other pressures for baroque design. To achieve economies of scale and a large order, an aircraft may be designed to be sold to both the air force and the navy (Kaldor, 1983, pp. 14–15). To satisfy the stipulations of both bodies, more and more features are added to the plane with the result that the final product is too elaborate, too costly and 'better' than either service requires. A similar process operates when international manufacturing collaboration is sought. Co–operation over the

European 'multi–role' combat aircraft, Kaldor claims, led to such elaboration that the plane 'cannot satisfactorily fulfil any of its roles, except perhaps the long–range nuclear mission' (p. 142). In one role it needs to be large for stable flight; in another it has to be small and manoeuvrable. Such developments mock the original intention of technical improvement.

The final aspect of military equipment which makes assessment of technical quality problematic is precisely the competitive nature of its task. This competition led to the initial constant demand for better and better weapons. But since betterness can finally be evaluated only in terms of the outcome against an opponent's weapons, assessments of betterness will depend on intelligence information and guesses about the opponent's weapons and the manner in which they will be used. Here again technical decisions are inseparable from judgements in which social variability is likely to be very great.

In this section we have seen seven sets of reasons why decisons about improvement in military technology cannot be simply technical. And these reasons do not stem from 'perversions' of technical assessment associated with bribery and politicking. They derive from complexities inherent in technical choice, many of which are analogous to the problems surrounding scientific choice discussed in Chapter 1. Just as the expense of weapons inhibits testing and training, the expense of building neutrino detectors leads to problems of replication. Just as there is doubt over how the speed of aircraft should be measured, there is uncertainty over the construction of a gravity wave detector. In both cases the technical or scientific evidence by itself is insufficient to allow conclusive decisions to be made. Accordingly, we can talk of the social construction of the decisions internal to the assessment of military technology.

The Sociology of Technical Choice

Both in Chapter 1 and earlier in this chapter, a distinction was made between sociological accounts which emphasized external influences on technical choice and accounts which extended sociological study to the technical criteria themselves. The study of the reaping machine manufacturer, for example, concentrated on the way in which political considerations overrode purely technical grounds for choice. Such a model for the study of technology effectively restricts the scope for sociological analysis. A discussion of military technology has allowed us to extend the scope beyond these restrictions. Even when weapons producers are aiming at the manufacture of better weapons there is room for negotiation over what 'better' means. Weapons manufacture clearly has vast political and economic implications. No doubt these pressures sometimes lead designs to be overlooked in favour of ones believed to be worse when the less good ones are more profitable or when they bring dividends in terms of international relations. But political considerations are not limited to occasions when the preferred alternative is refused; they enter into the actual constitution of the criteria for betterness. There are no purely 'internal' technical choices, since technical evaluations are suffused with political, economic and more broadly social considerations.

The point at issue here can be exemplified by briefly looking at two cases. The first, the decision over the Nimrod early warning radar system, has already been introduced. A statement announcing the cancellation of the Nimrod programme was made to Parliament on the afternoon of 18 December 1986. That evening MPs were given the opportunity for a three–hour discussion of the issue, during which the bases for the decision were reviewed. The Secretary of State for Defence had stated his decision clearly but had acknowledged (House of Commons, 1986b, p. 1350) that it was 'a question of scientific and engineering judgement ... I must judge the prospects on the basis of the scientific and service advice available to me'. The role played by 'judgement' was taken up by the first opposition spokesman, who sought to bring into question the government's ability to exercise good judgement. Other speakers joined in to suggest that the RAF had long been opposed to Nimrod and that it was unjust that RAF officials should be the assessors of the equipment (p. 1428). Thus, some opposition to the decision was built around the idea that the decision–makers were incompetent or that they had a vested interest in the outcome. Other critics, however, concentrated on the way in which the component elements of the judgement had been weighted. Thus James Prior (a government–party MP and chairman of the main avionics contractor, GEC) questioned the price calculation which had been used to justify the purchase of the Boeing–based equipment. He picked up on an earlier admission that eight Boeings would finally be needed, not simply the six already ordered, and argued that inflation had to be taken into account. Consequently, he argued, the deal for the United States equipment will demand 'precisely double the amount that the 11 Nimrods are due to cost and which we [GEC] have guaranteed. It is not £200 million more [the figure suggested earlier by the government spokesman] but more than double' (p. 1431).

For the opposition, Tam Dalyell made a related point, inquiring into the 'through–life costs' of the two systems (p. 1437); even if the Boeing system were cheaper to purchase, would it be cheaper overall? He also made the point that the US system was becoming aged. If the government allowed Nimrod to be finished the UK would have the more up–to–date system and one which might well be more secure in operation. On this view, it would be reasonable to gamble on the future of Nimrod, especially since the US product would anyway not be finally delivered until the 1990s. These objections and counter–arguments make the same point for technical choices as Kuhn made about the selection of scientific theories; people 'fully committed to the same list of criteria for choice may nevertheless reach different conclusions' (1977, p. 324). Even on the basis of internal criteria (cost efficiency, long–term viability and so on) and even in a case where the loser, Nimrod, was held up for ridicule, the technical decision is not self–evident. It is a matter of judgement and is thus inescapably suffused with, in this case, political and economic considerations.

The second case concerns the 'strategic defence initiative' (SDI). This scheme offers the prospect of a defensive shield, including space–mounted laser weapons, protecting the USA against in-coming nuclear missiles (B. Thompson, 1985). SDI has excited strikingly different receptions. Its advocates hold that it will finally make the USA secure against the Soviet threat since it offers a far better prospect for defence than the present system of nuclear retaliation and mutually assured destruction. If successful, it would simply make nuclear missiles redundant. Opponents argue that it could not be completely secure and would not, therefore, operate as a viable defence. Worse still in their view, it has the decided disadvantage that it takes conflict into space. Finally, there is concern that even if SDI is not effective as a defence against an all–out nuclear assault, it would be a useful in an offensive context. A reasonably effective defence would permit the USA to launch a 'first strike' attack itself, secure in the belief that the few remaining missiles with which the enemy could retaliate would be stopped by the SDI. The possession of a defensive capability thus means that a nuclear war could be fought, won and survived. For this reason, opponents of SDI regard it as an escalation of the arms race.

Great controversy has been stirred by the SDI proposals. In the early stages of the development of the ingredients of the weapons

141

systems, for example the X–ray laser (Mangold, 1987, p. 5), the situation was akin to the 'producer sovereignty' described by Galbraith (see Erickson, 1971, p. 228). A group of carefully selected young researchers worked under very intense conditions (ably captured in Broad, 1985) on the elements of a defensive system. The idea was then presented to President Reagan as the fulfilment of his strategic and political requirements (E. P. Thompson, 1985b, pp. 18–19). Opposition to the proposals was mobilized in terms both of the inadvisability of increasing weapons spending and of the inadequacy of the SDI technology. On this latter score it was argued that counter–measures would be too easy to devise, that the necessary accuracy and computing power demanded by SDI could not be guaranteed and that the defensive weapons would themselves be too vulnerable. SDI's supporters could accept the technical criticisms, at least in part, and argue that research was needed to meet the shortcomings. However, opponents claimed that the weaknesses could not all be eliminated. And this meant that SDI could have only an offensive purpose. The supposed technical imperfectibility was their evidence that all along the objective had been offence.

In the debate over SDI both sides have placed great emphasis on the technical and strategic issues. In this respect the dispute resembles the case of the phrenological controversy. Although moral, political *and* technical issues were at stake, the disputants chose to conduct much of the controversy at a technical level. But the issue could not be resolved by scientific and technical observations alone; the parties to that dispute saw the brain differently. The case is similar for SDI. Although both sides continually appeal to technical issues, these can never be decisive. What one side regards as insuperable difficulties, the other interprets as mere technical hitches. An experiment heralded by SDI's advocates as an important success (Broad, 1985, p. 106) is dismissed by its opponents as irrelevant to the actual conditions under which SDI would be used. The inconclusiveness is further encouraged by the fact that no defensive system can truly be tested except by a live attack.

Conclusion

This chapter has two levels of conclusion: about military technology and about technical improvement. The former are of considerable substantive importance but can be concisely summarized. While steps have been taken to decrease the coherence of the MIC and firms' dependence on the state, many companies are still heavily locked in, particularly in Europe (Smith and Smith, 1983, p. 73). Attempts have been made to introduce more competition into defence contracting, but continuing needs for secrecy and innovativeness mean that there are limits to accountability. The pressure on research and development to yield notional improvement is likely to continue to promote baroque designs. Moreover, there are good reasons for thinking that high levels of military research in the national economy have negative consequences. Such research absorbs a great deal of scientific and technical expertise. Second, the products of such work may be too specialized to be successful in the civil market. This is partly a question of the gold–plating which goes on in weapons design but it also arises from the discrepant demands of military and civil use. Military components are extremely specialized and are supposed to meet very high standards of durability and reliability. These considerations are at odds with the demands of general utility and cheapness which yield success in the mass market (for microprocessors for example). Even in the market for military goods over–specialization may be a handicap. Arms producers are relying more and more on overseas sales, often to Third World countries. This strategy has been encouraged by governments seeking to reduce companies' dependence. But it is by no means always the most intensively researched products which sell well (Gummett, 1984, p. 74; 1986, p. 61). Thus it is a reasonable supposition that defence spending will continue, as Smith and Smith observed, to correlate badly with economic success.

The study of military technology also imparts lessons about the nature of technological improvement. We have seen that social considerations enter into the evaluation of technical criteria even in the case of military technology where the direction of improvement would seem to be particularly straightforward. In Gorz's terms, the use–value of weapons ought to be quite clear. But in every case the estimation of technical utility is far from straightforward. It is bound up with social and, frequently, with political judgements, whether it is the speed of a fighter, the utility of Nimrod, or the strategic value of SDI. Judgements about these matters are as suffused with social considerations as were the judgements passed on gravity wave detectors or on the flux of solar neutrinos. The social constructionist approach developed in Chapters 1 and 2 can be helpfully applied to the understanding of technical change and the assessment of technical merit.

Finally, this conclusion allows us to pick up Habermas's point about the scientification of decision-making. Certainly, the case of military technology illustrates this issue well, since defensive decisions are generally made by a few people, who claim exclusive technical expertise. This claim allows them to exclude from the decision–making process people for whom defence decisions have huge practical implications. Habermas warns against this development since he believes that issues other than purely technical ones are at stake: for example, moral issues. The argument developed here allows Habermas's concern to be expressed more precisely. Because there is no purely internal technical logic, technical authority is bound to be imbued with political and social considerations. Scientification is a cause for anxiety not because it excludes forms of reasoning other than technical ones but because it purports to have screened off all non–technical factors when, as we have seen, even internal technical criteria have social elements.

Notes: Chapter 5

1   It should be acknowledged that there have been some influential voices raised in the USA asserting that further arms development is futile and dangerous. A book arguing this case has recently been published by Robert McNamara (1987), former president of the Ford Motor Company, World Bank president and US Defense Secretary. That people of influence in the USA can be found arguing for either side serves further to break down the credibility of Baran and Sweezy's talk of the US 'oligarchy' as though it had some unified existence.

2   The persistence of costs–plus agreements in the UK into 1985 was reported in an interview on BBC Radio Four's *World at One* broadcast on 22 December 1985. A figure of around $7^{1}/_{2}$ per cent was suggested as the usual allowance for profit. In the parliamentary debate over Nimrod a figure of 4 per cent was cited (House of Commons, 1986b, p. 1429).

3   This article is reprinted in a most useful collection by MacKenzie and Wajcman, 1985b; the editors also offer valuable comments on the social shaping of military technology, 1985b, pp. 226-31.

[6]

Technology, Science
and Development

Using Technology to Develop

In the Introduction we saw how the idea of development as modernization readily led to the proposal that underdeveloped countries should be assisted by allowing them access to modern technology. Since the West managed to develop even without the benefit of the technical skills now at hand, we might reasonably think that the route to prosperity for the poorer nations today would be made rather simpler if they were able to import this knowledge and the accompanying technical products. They could move from elementary technologies to modern, scientific ones in a very brief space of time and therefore shorten the period needed for the modernization of their society. However, after more than thirty years of technical assistance to underdeveloped countries the story is one of very partial success.

There are spectacular achievements: the construction of dams and hydroelectric power stations; the ownership of satellites; the availability of modern pharmaceutical products; the growth of complex infrastructural systems, such as electrical supply, telecommunications and roadways; and even the establishment of nuclear power programmes. Hand in hand with these achievements however there exist vast problems. The economic benefits associated with these technologies often accrue to a minority of the population. The 'modern' elements in the economy have frequently served to widen the gap between the wealthy and the poor, the powerful and the powerless. They encourage the development of a 'dual economy' with a divide between a wealthy, modern

145

sector and poor, subsistence areas. The resources which have been directed to acquiring the technologies could in many cases have been used to allay problems of unemployment, education or healthcare. The environmental consequences of new techniques have also been far from uniformly beneficial. Arguably, therefore, the benefits of the technology have entailed high costs for the society. And, at the same time, in some cases the prestigious new production technologies have themselves not worked optimally or even satisfactorily. Finally, the majority of the output of the Westernized aspects of the economy is usually not for home consumption; it is traded for further modern goods or for the foreign exchange required to purchase more plant, fuel and so on. The importation of technology for purposes of development has thus been bedevilled with problems; these problems are usually discussed in terms of the inappropriateness of Western technology and the formation of technological dependency.

Inappropriate Technology

Western technologies can be seen to be inappropriate for a very wide range of reasons. In some cases they may be inappropriate in a very literal and obvious sense. Thus, there are often great climatic and environmental differences between underdeveloped countries and the principal manufacturing nations, particularly of Western and Eastern Europe. Products which are suitable in the countries of manufacture may simply not fit the country to which they are exported. In 1985 the UK government was keen to encourage a deal whereby helicopters worth £65 million would be purchased from the Westland Company by India for use, for example, in the oil industry off Bombay. Given the subsequent, well publicized difficulties experienced by this company one may feel that there were mixed motives behind the encouragement given to this sale. But leaving aside such issues for the moment, one of the grounds for Indian reluctance in this arrangement was a reported unsuitability of the helicopters for the extremely hot conditions which commonly prevail around Bombay (*The Times*, 3 May 1985, p. 1). On this view, the technical product was not appropriate to the intended context. Given the enormous geographical variation between the underdeveloped countries, technologies may suffer because the environment is too hot or too cold. There may be problems of adaptation to other aspects of the environment also;

146

in mountainous countries the air may be thinner (this too may, for example, pose a problem for helicopters); the environment may be dusty or particularly wet. Such brute considerations may affect the actual efficiency or suitability of machines designed for European or North American conditions.

From a European perspective countries which experience these difficulties may seem to be faced with natural extremes. We tend to regard our design as corresponding to the 'normal' occasion and put any practical failure down to the 'abnormality' of the area to which the products are being introduced. On reflection, though, this indicates simply that technologies can be evaluated only according to the assumed conditions of use. And there are rather particular circumstances for which our technologies are commonly designed, conditions which may seem normal to Europeans but look distinctly abnormal elsewhere. The majority of motor cars, for example, have low ground clearances. This is related both to issues of stability and aerodynamics and to stylistic considerations. Such a design functions satisfactorily on the road surfaces of the richer countries. It makes a lot less sense on roads which are poorly maintained, or which are annually rutted by rains. Thus apparently technical questions about good design or adequate construction are highly relative to assumptions about conditions of use and even to considerations of fashionability.

However, the utility of technical products is not only relative to issues of physical circumstance. Appropriateness also has to be assessed in the light of the cost and accessibility of the other inputs which go along with the technology and of the availability of possible substitutes for Western techniques. For example, changes in production technologies have commonly been related to the cost and availability of the raw materials on which they are based or which they consume. The French move towards nuclear power in electricity generation over recent years has been assisted by the steep rise in the cost of imported fuel oil. The nations with expertise in new technologies may then be interested in trading them with underdeveloped countries. France and the UK have recently negotiated involvement in the establishment of nuclear power facilities in China (*The Guardian*, 18 July 1986, p. 18). But there is no guarantee that the conditions faced by the Western nation are the same as those confronting the underdeveloped world. Yet technical advance and the design of innovative products are stipulated by the situation of Western countries. What, for them, appears to be the best may not suit the circumstances of their potential trading partner. In the same way, apparently less good techniques may be of value outside the West. Steam railways, for example, may be appropriate to underdeveloped countries where wood is available or where there is access to coal, especially if they do not wish to spend currency reserves on importing oil and when the expensive electrification of rail transport is not practicable.

This difference in the pricing of inputs has a further and particularly acute implication. One of the inputs considered and priced alongside others by economists is the cost of labour. In the richer, capitalistic countries, where manufacturing investment decisions are largely in the hands of private factory owners and directors, technical improvements have increasingly been aimed at the reduction of labour costs. We have virtually come to accept that 'labour saving' means 'improved'. Productive technologies have increasingly been directed to reducing the human contribution to manufacturing. This has meant the substitution of cheap inorganic energy (from electricity or fossil fuels) for human effort. It has also frequently had the effect of decreasing the dependence of manufacturers on large numbers of employees and of substituting complex machines for the specialized skills of workers. These effects have had some influence in shifting the balance of power in industry between the employed and the employers. In most cases this question of control has largely been incidental or a welcome by–product of the drive to cut costs (although the case of moulding machines described earlier should be borne in mind here). Workers who have been relieved of their jobs have either moved to other employers or joined the unemployed. Particularly at times of growth this has been a successful strategy. Increases in the productivity of each worker have gone hand in hand with a general increase in prosperity. Such a strategy has even been mimicked by state socialist countries where managers have been encouraged to run enterprises on a more individualized and capitalistic model. Indeed, the Hungarian authorities have recently accepted the recognition of unemployment as a step in increasing workers' productivity (*The Independent*, 4 July 1987, p. 8). Surplus labour can be identified and personnel switched to new tasks or to areas in which productivity cannot be increased by mechanization and automation. In any case, the consequence of this tendency has been for an increase in capital intensiveness to be generally equated with betterness. The apparently natural direction of tech-nical change has been away from labour–intensive technologies to ones depending on high levels of investment in plant and machinery.

When the countries of the West, but also the East, have offered their best technology to underdeveloped nations it has usually meant best in terms of the criteria prevailing in the producing nation. Accordingly, countries with very little industrialization have been offered technologies which are capital intensive (and therefore expensive) and which are dependent on refined and costly inputs but which offer few employment opportunities. The acceptance of such a route to technical modernity has thus presented the dilemma that it is not accompanied by large employment opportunities. The result is modernization straight into high levels of worklessness. This problem is further compounded by the fact that such capital–intensive plant generally requires highly trained operatives and specialized maintenance. Worse still the acquisition of this technology places a great burden on the foreign currency resources of the underdeveloped countries (although as we shall see later on this may be lessened by various forms of aid). Moreover, capital which is directed to the purchase of foreign technology is no longer available for investment in indigenous industry or in research and development. For these reasons there is a danger that foreign technology will be inappropriate for the demands of the workforce, for the national balance of payments and for the level of support for domestic manufacturing enterprises.

To summarize, even if the attempt is made to offer the best of current technology to the underdeveloped world it would risk inappropriateness in five ways:

(1) It might be unsuited to the conditions and environment.
(2) It would be capital intensive and thus expensive. It would tend to be costly both to acquire and to run. It would absorb financial resources to the probable detriment of domestic industry.
(3) It would be dependent on fossil fuels or modern forms of power and on refined or processed inputs.
(4) It would create few employment opportunities when underemployed labour is abundant; it might even displace local people who are making a similar product by traditional means.
(5) Its maintenance and operation would require skills and equipment unlikely to be found in the receiving country.

Technology Transfer and Technological Dependency

Up to this point we have examined the difficulties which have arisen in attempts to transfer Western technologies to underdeveloped countries. The inappropriateness of the technology has meant that technology transfer has not set the poorer countries on the modernizing path pioneered by the West. Some analysts, however, would put the case more strongly (Cooper, 1973, pp. 8–16; Stewart, 1978, pp. 114–22). They suggest that the transfer of inappropriate technology has actually increased the underdeveloped world's dependence on the West (and, to a lesser extent, the East). According to this argument, a crucial issue in the transfer of technology is the question of control over the technology. In addition to its possible unsuitability, introduced Western technology can often bring the drawback of a loss of control over parts of the domestic economy, since the equipment may be under the control of a foreign company or of local people whose interests are allied with foreign firms. In short, the transfer of technology may suit the interests of those doing the transferring more than the needs of the underdeveloped economy.

To understand the idea of dependence being used here we should briefly return to the discussion of social development provided in the Introduction. In that discusssion the point was made that Third World countries are not now in the same situation as were the European countries early in their industrialization. In particular, where the European countries had colonies whose resources they could exploit, the nations of the Third World have themselves typically been colonies. They are not simply undeveloped; they have been exploited or underdeveloped. On this interpretation, the economic condition of underdeveloped countries cannot be understood without referring to the world economic system in which they have been involved. During the colonial period, but also since formal independence, their economies have been shaped by their dependence on the 'core', industrialized nations. Specifically, this means that their economic institutions and infrastructure have been structured in response to the demands of the core nations. For example, land has been organized for the growing of crops, like rubber, cotton, and special timbers, as well as plantation foodstuffs, which provide the raw material for Western industries. Infrastructural building has been designed to facilitate the extraction of these raw materials from the countryside. At the same time, processing and manufacturing development has not been encouraged. A division of labour has been practised with the dependent regions providing raw materials in exchange for the finished goods produced in the core.

The recognition of this exploitative relationship and the client status which it has imposed on underdeveloped countries has been at the heart of dependency theory (Palma, 1981, pp. 42–62; Roxborough, 1979, pp. 60–9). Without elaborating on the differences of detail which characterize the various dependency theorists, it is possible to state that they oppose theories of modernization and argue that relations between the First and Third Worlds have very little to do with encouraging development and a great deal to do with continuing the subjugation of the Third World to the core. The very steps which have been taken in the underdeveloped countries to encourage extractive industries and the provision of raw materials serve to maintain their dependence on the industrialized nations. The economic structure of dependent countries is generally antithetical to independent development.

This is not to say that no one in underdeveloped countries benefits from the attention of the First World. Some colonial powers encouraged the collaboration of local leaders and landowners, who profited greatly from the sale of their countries' vegetable and mineral resources. Similarly, after decolonization local leaders often took on the estates and prerogatives of the colonial power. Commonly an enclave of comparative wealth arose around a port or trading centre. Thus a powerful minority of the population was aware that its material interests lay with the continuation of the dependent relationship. It could be said that they exploited their hinterland just as the core nations exploited the colony.

When technology transfer is considered in this context it is easy to see that the technology might well compound the problems of dependency. Thus, the new techniques would tend to get adopted in the wealthier enclaves (the prosperous part of the dual economy) and would therefore serve to exaggerate differences between urban and rural areas. Only the relatively wealthy would be able to gain access to the fruits of the new techniques. The enclaves thus become islands of modernity. The technology could also be used to extend the international division of labour between the wealthy countries from which the technology originates and the underdeveloped countries which act as machine minders.

The impact of transfers of technology can be assessed only by considering the context in which the transfers actually take place. A major element in setting this context is the nature of the agencies which carry it through. A great deal of technology transfer is performed by large Western companies setting up subsidiaries or semi–independent companies or participating in joint arrangements with local capital. Such companies might assemble products, like motor cars, for the regional market or manufacture pharmaceuticals under licence. They might, on the other hand, manufacture a part of a larger product. At one level we would expect underdeveloped countries which engage in such arrangements to enjoy a favourable position since they can offer cheap labour and often command natural advantages such as proximity to the source of raw materials. Enterprises established in these areas should benefit from their ability to undercut the prices charged by producers operating in the high–wage, industrialized economies. Gradually, as profits and indigenous skills build up, these countries should be able further to develop their own industries, and their level of wealth should increase. Advocates of the free market suggest that this is the best way for economic development to occur.[1] Low prices will attract rational investment from all Western countries; no calculating business person will esteem national loyalty over a sound investment, so wealth will gradually trickle away from the USA, France, the UK and so on to the emerging economies until their increasing wealth forces up wage levels and turns investment away. Supporters of this position are able to point to the profitable enterprises of East Asia: South Korea, Taiwan, Hong Kong and Singapore (Berger, 1987, pp. 127–32; Bienefeld, 1981; see also Smith *et al.*, 1985).

This argument can certainly not be ignored. The success of these Asian economies (sometimes known as newly industrializing countries or NICs), together with other disagreements which divide dependency theorists, have led to increased uncertainty and flexibility in development theory (Fransman, 1986, pp. 59–64). None the less, these success stories are by no means widespread, and the symptoms of a dependent economy are still common on a worldwide scale. At this stage in the debate it is unclear which factor stands in need of special explanation: the NICs' apparent escape from dependency or the perpetuation of dependence elsewhere. But whichever pattern is seen as the exception and which as the rule, we can advance a number of reasons for the continuation of dependency.

There are, first of all, changes in the world economic climate. During periods of global recession countries will be eagerly looking for large companies with capital to invest. Potential hosts will therefore compete for the opportunity to bring in wealth and economic activity. They are tempted to make seductive offers to potential investors. For example, some help may be offered with capital costs; a reduced rate of taxation may be proposed for the initial years; employers may be excused obligations to employees (such as medical care or insurance); and governments may undertake to reduce the likelihood of disruption by trades unions. There is a further cost here also because local capital will not go unaffected by these investment patterns. Local investors will see more security in throwing their money in with a large internationally renowned investor than in supporting a local enterprise. Thus the welcoming of these companies may disrupt indigenous development. When these incentives are provided the short–term benefit to the host economy may be very small. As was mentioned in Chapter 4, even peripheral European countries, such as the Irish Republic, found this to be the case. Immediate advantage for the host country is sacrificed to the prospect of long–term investment and the creation of enduring employment. In theory the wooed investors will bring advantages to the economy in the long run. But other potential hosts may be aiming to attract these investors away before the long run arrives. By playing potential hosts off against each other companies may be able to move their investments around, maintaining high levels of profitability and with minimum costs. Local politicians keen to record short–term successes may make advantageous–looking deals; it is easier to attract investment than to retain it. The anticipated pay–off, whilst available in principle, may just never be realized. Thus 'investment' in an underdeveloped country need not entail any real transfer of economic activity; it can readily encourage further dependency and exploitation.

A second complicating factor is the operation of local self–interest. The image of investment discussed so far depicts a financial benefit to the local economy. But as the widely reported success of ex–President Marcos of the Philippines in amassing a personal fortune indicates, the ultimate destination of capital may be greatly at odds with local economic needs. Just as we saw earlier that inducements are sometimes used to secure deals within the military-industrial complex, so also can gifts and bribes be influential in technology transfer and investment decisions. Part of the supposed investment can disappear into the pockets of the local élite who do not share the economic interests of the nation as a whole. Regional interest may differ from national interests, and self–interest differ most of all. Corruption will inevitably decrease the capacity of the technical investment to operate for the good of the economy.

The third factor encouraging the persistence of dependency is the strategy of investing firms themselves. Hoogvelt (1982, pp. 66–72) suggests that there have been three general phases in the relationship between investing companies and underdeveloped countries. Initially, foreign–owned multinational enterprises controlled their plants in underdeveloped countries through shared or complete ownership. They transferred their technology into their own subsidiaries and, frequently, repatriated the profits to their domestic base. However, local governments realized that they gained rather little from these profits and that they had next to no say over what was produced in the factories they were hosting. In an attempt to alter this situation, the host countries began to demand a considerable stake in the ownership of companies. According to Adejugbe (1984, p. 577), 'the decade of the 1970s' saw this demand spread throughout Africa. Since the controllers of foreign capital were keen to continue their profitable business they started to submit to their hosts' demands for ownership. But this entailed little sacrifice on their part since the multinational companies were able to continue to exercise control through their rights over the product. By using their control over the patent they could regulate the amount of the product which could be made and stipulate where it could or could not be sold. And, if through the ending of ownership they lost direct control over the profits, this could be compensated in a number of ways. The parent company could make charges for production licences and could impose high prices for patented inputs. Instead of openly extracting profit, companies can disguise it through 'transfer pricing' – the placing of high charges on transfers of goods and know–how.

Host countries soon became aware of this new method for retaining effective control and sought to institute contracts which limited the scope for such operations as transfer pricing. But the companies responded quickly and, as Hoogvelt (1982, p. 69) notes, devised a new practice,

> involving the relocation of control over their foreign operations away from 'legal' ownership of the product, and towards the technological properties of that product. By making the physical characteristics of their plant and machinery, of their technical operation, and of their end–product, critically different from other similar machinery, processes and products that are available in the world markets, *MNCs can establish and preserve future supply, servicing and maintenance links quite*

independently of any written agreement or any form of legal ownership [italics added].

Some of the ramifications of this latest practice can be detected from Hoogvelt's study of firms in Nigeria which had been transferred from foreign ownership to local, private control. In these 'indigenized' companies she found that foreign capitalists rapidly moved to technological control. They quickly passed over the stage of 'legal' controls since 'recent political changes ... made foreign investors reluctant to place much faith in legal agreements' (1980, p. 262). Much of the business of these indigenized firms bore the signs of inappropriateness. They rarely stimulated other local industry to supply their needs (p. 267), they depended on processed inputs (p. 265), they were not labour intensive (p. 271), and the indigenization had brought a stake in ownership to only very few élite Nigerians (pp. 259–60). Ironically, the technology supplier's reliance on technological control had exacerbated the problems of inappropriateness. To ensure the exercise of control, technology of the highest specification was preferred (p. 263). Similarly, to avoid penalties on the repatriation of profits from the business itself, the import of highly processed inputs is encouraged to maximize the trading profits (pp. 265–6). Thus the problems of inappropriate technology and technological dependency may actually be increasing in areas outside the NICs with the growth of technological dependence as a management strategy for retaining control. The more crucial that control over technology becomes the lower the likelihood that technology and the access to technical skills will be freely transferred.

Dependent Science?

We have so far examined a number of reasons for the technological asymmetry between the industrialized and the underdeveloped world. Dependence on Western technology may be cultivated as part of a strategy of control. Equally, technical innovation generally demands great financial resources, which are concentrated in the developed nations. A third factor is the lack of scientific and technical skill and training in underdeveloped countries. According to figures for 1980 supplied by UNESCO (1986, p. V 17), there were only about 127 scientists and engineers actively engaged in research and development per million of the population in underdeveloped

155

countries. The comparable figure for the developed countries was nearly three thousand. And this is the average figure; the number for Africa (excluding the Arab states) is only 49 per million. For Japan the number is nearly 4,500 (p. V 14), well over ninety times the African density.

These figures for scientific and engineering staff are matched by the details of spending on research. As was mentioned in the Introduction there is a general association between a country's wealth and the amount of resources it is able to devote to science and technology. The relationship is not however a simple linear one. By and large, the richer the country the higher the proportion of its wealth it is able to expend on science and technology. Whereas Sweden, Japan, the USA and West Germany spend around 2.5 per cent of their great wealth on research and development, Pakistan allocated only around one–fifth of one per cent in 1979, a smaller proportion than in 1969 (UNESCO, 1986, pp. V 14–V 18). The 1980 average figure for the underdeveloped countries was 0.45 per cent.

Given the general connection between technical ability and the creation of industrial wealth these figures are clearly a cause for concern. They suggest an inequality of scientific and technical resources which is even greater than the prevailing disparities of wealth. The extent of this inequality cannot be attributed to the West's 'monopoly' of scientific thinking. Admittedly, the kind of scientific developments described in Chapter 2 are largely restricted to Europe and North America but as Rosenberg (1982, p. 245) has pointed out the 'three great mechanical inventions – printing, gunpowder, and the compass' identified as most influential at the time of the scientific revolution derived from outside Europe. Non–Western societies long led European developments not only in specific technical feats but also in experimentation, in conceptual development and in the study of health and healing (Goonatilake, 1984, pp. 6–14). It can reasonably be argued that non–Western science and technology have been victims of underdevelopment, particularly in the cases of China, South Asia and, spectacularly, Central America.

If we turn to the current situation of science and technology in the underdeveloped world we find it relatively little studied. The evidence which is available confirms many of the points raised in Chapters 3 and 4. Thus, there is no direct connection between the amount spent on science and its economic or social utility. Indeed, the scientific community in Third World countries is often spoken of as 'alienated' (Clark, 1980, p. 78). In this context this term refers to the fact that scientific workers are peculiarly cut off from any economic and cultural influence. As though acknowledging its lack of utility, the scientific community is often highly concentrated in basic science:

> In advanced countries far more is invested in applied and development research than in basic research. The ratio of these expenditures is 9 to 1 in France, England and the United States and about 4 to 1 in the rest of Western Europe [in the 1960s]. In Latin America on the other hand, these ratios are more like 1 to 4 or 1 to 9.
>
> (Herrera, 1973, p. 22)

As such the typical scientist is not even aiming to work on problems of direct local or regional significance. Furthermore the agenda for basic research is, so to speak, largely set by the Western countries if only because that is where the majority of it is done. Goonatilake describes the Third World basic scientist as 'always looking to an invisible jury abroad' (1984, p. 143). Often the scientific community is not large enough to support many workers in each area of science, which may lead to feelings of isolation. As one respondent to a survey of Indian scientists expressed it, 'our excitement comes by mail from outside. It depends on the postal system. This is the worst part; the spirit is dead' (cited in Goonatilake, 1984, p. 104; see also Choudhuri, 1985, pp. 479–82). And although the connection between basic research and technical advance is very indirect, there is inevitably a greater articulation between Western science and Western technology than between Western science (wherever it is practised) and non–Western technology. This observation about articulation is not restricted to those instances in which a technology can be developed from a scientific idea. It applies equally to the kinds of informal contact between scientists, technologists and industrialists which, in Chapter 4, were seen to be so important. Without these connections scientific activity may be a net drain on the economy: as Cooper (1973, p. 5) states, 'science in underdeveloped countries is largely a consumption item, whereas in industrialized countries it is an investment item'.

Overall, therefore, the scientific establishment in underdeveloped countries is poorly adapted to meeting the problems posed by inappropriate technology and technical dependence. Where there is great scientific skill it is likely to be directed towards goals stipulated by the international research community, itself based on the West. Furthermore, contemporary basic science is generally

157

capital intensive, a fact which further penalizes Third World scientists and encourages a 'brain drain' (Vessuri, 1986, pp. 28–31). Ziman (1981, p. 20) notes the irony that Third World scientists were able to excel at one of the most esoteric of the sciences, theoretical physics, because of its low demands for capital equipment: until the rise of widespread reliance on vast computing power, theoretical physics demanded only pencil and paper. From the point of view of industry, on the other hand, there is likely to be little encouragement for local scientists. Introduced industry has little need of local research and development since these functions can be performed by the parent company in the West, while the technical sophistication of local industry is unlikely to match the aspirations of the most skilled scientists. In this way there is a lack of the kind of positive interaction between scientific skills and technical needs which can occur, at least sometimes, in the West (Yearley, 1987b, pp. 199–205).

Aid, Technology and Dependency

As commercial transfers of technology have distinct drawbacks and underdeveloped countries are low in indigenous scientific and technical capability, an important avenue for technical assistance is provided by inter–governmental assistance or aid. Aid has come very much to public attention in the mid–1980s as a result of the help brought to Ethiopia and the Sudan by aid agencies such as Oxfam and Save the Children. However, the great majority of aid is not dispensed through such non–governmental organizations (NGOs) but takes the form of grants or very low–interest ('soft') loans from one government to another. Aid of this sort, more technically known as Official Development Assistance, is not generally speaking directed to the relief of emergencies. It is mostly concerned with fostering development projects, and one prominent way of achieving this is to provide financial aid for the purchase of technical equipment which the underdeveloped countries lack.

Aid in this sense refers to vast sums of money; in 1981 it totalled nearly 36 billion US dollars (Hayter and Watson, 1985, p. 9). Most analysts of aid distinguish between bilateral and multilateral aid. The former refers to assistance offered by one national government direct to another. Aid from the USA to Egypt or from the UK to India would fall into this category.

Multilateral aid is assistance offered by agencies acting for groups of nations. For example, the European Community has an aid agency which directs grants and soft loans to poor countries. According to Hayter and Watson (1985, p. 13), over three–quarters of the total sum is distributed in the form of bilateral arrangements. Of this sum only around 12 per cent is dedicated to food aid and disaster assistance. On the other hand over half is provided for the sponsorship of particular development schemes. These schemes are commonly for construction, infrastructural projects or industrial development. They are therefore major contributors to the import of technology.

Since this transfer of technology is said to be motivated by a desire to help the recipients' economic development, one might anticipate that it would be plainly beneficial. However, many critics assert that this is not the case. Although aid represents only a very small proportion of the spending of governments in industrialized countries it does not escape from considerations of political interest. Minimally, aid is directed in such a way as to reflect well on the donor and is generally arranged to assist the donor's foreign policy objectives. Often, as we shall see below, aid is related quite directly to the commercial interests of the giver. For these and similar reasons, much criticism has recently been directed at Official Development Assistance.

Turning to the particular connection between aid and technology, three principal objections have been raised to assisted technical transfers. First of all, high–technology projects may not be fully funded by the aid. Accordingly the assistance may commit the recipient nation to expense which it may not be able to afford or which it would not have chosen had there not been a 'special offer'. By absorbing some local capital the aided project may act to divert indigenous investment away from local projects. As Hayter and Watson suggest (1985, p. 13), this aid

> usually finances part of the total cost of the project, often only the direct foreign exchange expenditures required in its initial phases. Thus it may finance the machinery, materials and skills which it is considered need to be imported from abroad for the construction of, for example, a dam, a railway, a power station, harbour, hospital or school; but it will not finance the foreign exchange expenditures required to keep it running, including expenditures on spare parts for imported machinery.

159

Short–term gains may be bought at the cost of long–term commitments. Sadly, as mentioned earlier, short–term considerations
are often uppermost in political leaders' minds.

Second, the technical projects which First World governments
choose to fund may be inappropriate in some or all of the ways
described at the beginning of this chapter. This may arise even
when technical transfer is carried out with the best will in the world.
However, aid projects cannot often escape considerations of political
interest. Hence, for example, a donor country may well be willing to
support the construction of a dam since this will be a very tangible
accomplishment which reflects favourably on the donor as well as
assisting in irrigation and electrical power generation. The donor
may even feel that such assistance is mutually beneficial since it will
provide work for the donor country's industry. None the less what
the donor chooses to fund may not be the most appropriate for local
requirements. For the same costs as are entailed by the construction
of one large dam, a great number of smaller irrigation schemes
could be financed. They could bring agricultural relief to a far
larger region and would not be associated with the environmental
problems with which dams are frequently linked. The virtues of
alternative schemes clearly cannot be decided without detailed
assessment, and I am not opposing dam construction *per se*.
The point is that Western aid arrangements are often predisposed
towards large–scale, high–technology projects. Such projects may
not create many jobs locally, although employment consequences
in the West may be favourable. Lastly, such schemes often commit
the recipient to continued expenditure, maintenance agreements
and so on. The additional argument is sometimes made that, since
such assistance tends to favour schemes like power production
and the establishment of railways and harbours, these can be seen
as laying the foundation for continued Western investment and
trading opportunities. In other words, donor nations are in some
sense underwriting the viability of subsequent private investment
from the West and predisposing the economies to development
along a Westernized and technically dependent path.

The final difficulty facing the aid–backed transfer of technology
stems from the practice, common in bilateral aid agreements, of
tying aid to the products and services of the donor nation. The
extent to which countries adopt this procedure varies although it
is common; on average around 50 per cent of bilateral aid is tied,
although in the British case the figure recently reached 70 per cent
(Hayter and Watson, 1985, p. 15). In practical terms this means that
project development assistance is commonly a way for Western

governments to support their domestic industry by providing a sales contract which, without the 'aid' subsidy, would not have been won. Little of the notional sum of aid ends up in the recipient country, since the money is paid directly back to an engineering or other contracting firm. If the aid is not in the form of a grant but is a low-interest loan, or if the project is only partially supported by aid, it may finally result in a net cost to the recipient. Of course the recipient does get the goods. However those goods may benefit only a limited part of that country's population or even be of most use to foreign investors. Sometimes the goods supplied are virtually 'bin ends': old or unpopular lines which, without the subsidy, would have needed to be heavily discounted. So although aid worth many millions of dollars may be given, the recipient is not free to spend that money how it wishes, nor to 'shop around' for the best bargain, nor to direct its spending in its own domestic market. Aid is a gift but often an expensive and inappropriate one.

With the economic hardships faced by all nations since the recession of the late 1970s, Western countries are becoming more explicit about the fact that aid must benefit the donor as well as the recipient. The UK, for example, has established an 'Aid for Trade Provision' (ATP) according to which a proportion of British aid is directed to projects designed to help UK industry.[2] According to Tanner, the judgement of the Department of Trade and Industry comes before that of the Overseas Development Administration (ODA) when ATP funds are sought (1984, p. 25):

> The ATP overtly subsidises British companies trying to win orders in the Third World, with the development value of the project coming in only as a second criterion ... Time is often short and the ODA examination can be 'at best perfunctory,' according to the Independent Group on British Aid.

If British trading interests are the first consideration, the danger that the subsidized technology will not be the most appropriate to the recipient nation is clearly magnified. Yet when recipients are faced with the choice between tied aid and no aid it is very hard to turn aid down. In any case such projects may well favour the élite urban areas where government officials live. The pursuit of self-interest is not confined to the First World. Still, it is Hayter and Watson's conclusion that in the long run underdeveloped countries would be better off without a great deal of aid as currently organized.

Two brief cases usefully exemplify the issues at stake here. As mentioned earlier in this chapter, part of the UK's recent aid to

India has taken the form of Westland helicopters. The £65 million of aid was not spent in India; it returned to the helicopter company and was thus effectively a subsidy to a troubled company. Of course, India did get the helicopters – although Tanner suggests (1984, p. 25) that they might have preferred those from the French firm Aerospatiale. But they were paid for out of a British budget for *aid*. In the future, moreover, the machines will need spare parts and maintenance which will bring Indian business to Britain. A second illustrative example was taken from *The Observer* newspaper by Webster (1984, p. 162). Assistance from the UK to the value of £150 million secured orders for British equipment for work on Cairo's sewers worth £500 million. In this case there is no data on the appropriateness of the British engineering equipment. However, it shows the logic whereby partial aid acts as an inducement to buy British. Aid is, so to speak, a 'loss leader'.

Overall, therefore, much First World aid is provided with at least half an eye to the donor's interest. Ths interest is intimately related to the maintenance of its industrial capacity and is thus almost certain to lead to the transfer of modern, capital–intensive and frequently inappropriate technology. In this way, aid tends to reinforce the processes creating a dual economy in underdeveloped countries with a marked division between the rich and poor, the modern enclaves and the underdeveloped hinterland.

A final and very significant consideration in the assessment of aid concerns the role of Western banks in advancing loans to the underdeveloped world. In addition to the money which flows between countries through Official Development Assistance, the 1970s witnessed an enormous increase in international lending by commercial Western banks. Effectively, these banks advanced loans to selected Third World countries as a way of investing the oil wealth of OPEC (Lever and Huhne, 1985, p. 13). They appeared to be doing everyone a favour; the banks themselves profited, Third World countries received capital which they could use as their rulers wished, and the oil–rich countries looked set to receive a good return on their money, since rapid, profitable industrialization of Third World economies was anticipated.

Towards the end of the 1970s the system began to look less than perfect. While interest was due on all the loans, not all of the capital was producing high returns. Sometimes this was due to mismanagement; sometimes the banks had been keener to secure a deal than to inquire into the uses to which the money was going to be put. Generally speaking, private banks are not as well designed as aid agencies for development financing. They are less able to ensure that finance is used fruitfully and they cannot readily soften the terms of their loans in response to changes in the world economy without endangering confidence in their own business. These weaknesses began to show clearly when dollar interest rates rose in the recession at the end of the 1970s. The interest charges on underdeveloped countries increased, while the general economic climate ensured that even those loans which had been well invested were performing less well because of an overall downturn in business. The well-known consequence is the 'debt crisis' of the 1980s. Many of the biggest debtors, particularly in Latin America, are struggling simply to meet the interest payments; they are quite unable to repay any capital. In some cases, such as Argentina and Chile, the interest payments alone each year are equivalent to about half of the country's export earnings (Hayter and Watson, 1985, p. 25). With the inevitable reduction in further lending, a situation has now arisen in which the underdeveloped countries are effectively handing over money to the industrialized world. As Lever and Huhne drolly note, 'no economist has yet advocated a large flow of resources from the poorer countries as a way of stimulating their economic progress' (1985, p. 11).

The debt crisis raises vast problems quite beyond the scope of this book. However, it can be stated that the loan–backed industrialization of the 1970s encouraged underdeveloped countries to import Western manufacturing technology. In this regard its consequences are similar to those of aid–supported technology transfers. With capital available on a scale never known before it was obviously very tempting to adopt a highly technological approach to development. As further funding dries up or is absorbed in meeting interest repayments, problems of technological inappropriateness and dependency can be expected to appear. Furthermore, the fact that there is now a net flow of finance to the First World indicates that aid is not such a gift after all. We will return to the issue of debt in the conclusion to this chapter.

Aid and Technical Solutions: the Green Revolution

Up to this point aid has been taken to mean financial assistance. The idea has been to counter the lack of development in the Third World by offering gifts of existing modern technology. However, attempts to help the underdeveloped world can also be made through the preparation of technical solutions to its problems. Probably the most famous example of this, and also a good illustrative case, is the 'green revolution' in agriculture. This term describes the introduction to general use of new, 'high–yielding' varieties of food grains developed in plant breeding centres. The original idea, on which research began in the 1940s, was that a solution to the problem of hunger in the Third World could be achieved if farmers were able to produce more food from the existing land (Fitzgerald, 1986). This could be accomplished by breeding varieties of wheat and rice which yielded more edible grain per plant and had a shorter maturation period so that more crops could be cultivated each year. New varieties prepared did display large measured increases in yield under trial conditions. However, to generate these yields, they required a regular supply of water and the application of agrochemicals: fertilizers, weedkillers and pesticides.

Supported by the US foundations which had financed the research, this proposed solution to the problems of hunger was widely adopted in underdeveloped countries, notably in India, Pakistan, Mexico, the Philippines, Afghanistan and Sri Lanka (George, 1977, p. 115). Great increases in food production were recorded in many areas, but the introduction of this agricultural technology did not, as seems to have been anticipated, just increase everyone's ability to produce food. Instead it can be seen to have had enormous social and political consequences, changing the balance of wealth and power in the rural areas and affecting the international trade and technological dependence of the importing nations (Byres, Crow and Ho, 1983, pp. 38–42). These changes associated with the green revolution bear out the main point made in Chapter 5. We cannot say that a neutral technology happened to have such–and–such social consequences; social patterns were already embodied in the design and evaluation of the technology. For example, the reliance of the new varieties on inputs of agrochemicals demands that growers trade some of their produce. Previously, many peasant farmers had grown food almost exclusively for their family's consumption. The very design of the technology demands that they change their mode of production. Equally, on a national scale, the entire country's food productivity now depends on the provision of agrochemicals. Often these are produced by foreign companies (George, 1977, p. 116). Thus, the food needs of the whole nation, often regarded as a strategic issue, come partially under the control of outsiders.

A great many studies of the green revolution have been undertaken (see Byres, Crow and Ho, 1983; George, 1977; Pearse, 1980), and the discussion here will inevitably be incomplete. A few of the most outstanding issues can, however, be addressed. The first of these has already been mentioned: the need, even of the poorest peasants, to buy inputs. Using traditional methods poor farmers can be virtually self–sufficient. One season's produce yields food and new seeds for the following year, while the crops are weeded and tended by hand. With the new seeds the same farmer ought to be able to produce a surplus each year which can then be traded for the necessary inputs and fresh seeds. The weeds are reduced by agrochemicals. However, the need to enter into trade holds perils for the small farming family. They have no control over the market price of the inputs or of the grain they wish to sell. Some years they may do well out of the prices and be able to buy extra food or new goods. Other years their income may not match their necessary expenditure. They will have to borrow (at exorbitant rates) to buy all the required inputs. Or they will economize and omit some of the agrochemicals, but then the new varieties will no longer give high yields. Whereas the traditional system guaranteed struggle and hardship, the new system obliges farmers to flirt with disaster. After one lean year or a period of low grain prices, they will face high rates of interest or successive poor crops. These are likely to lead the family to sell some of their land, even to become landless.

In this way the green revolution has seen great changes occur in the economic structure of rural areas. Poor families have lost land to the relatively wealthy. And this has not happened only through the mechanism above. The process has been assisted by economies of scale. For example, although all farmers need water, wells have a minimum cost. They are thus more readily available to the relatively wealthy (Bell, 1973, pp. 141–2). Equally, the larger farmers have better access to credit and on less punitive terms. In this way the regions in which there has been a successful green revolution have witnessed a polarization of the social structure. Those previously relatively wealthy have tended to gain land and to achieve a cash profit, while the poorer families have tended to lose land. They have had to seek work as labourers on the farms of the wealthy and become, so to speak, a rural proletariat. Furthermore, if we take the case of India, regional inequalities have been reinforced. Wheat and rice are not grown everywhere. Poorer areas tended to grow crops such as sorghum and millet which had already been selected because they could withstand harsher conditions. These crops were little touched by the research on high-yielding varieties. The wealth of those who grow them has fallen behind the wealth of successful green revolution farmers. It is also suggested that the new technology has adverse dietary consequences for those farmers who live exclusively off their own produce. Untreated fields of grain used to contain, among the weeds, occasional protein–yielding plants which were of dietary value; agrochemicals ensure that green revolution crops are monocultured (George, 1977, pp. 122–3).

A further stage in the polarization of the rural areas has seen the wealthier farmers introducing tractors and even combine harvesters. Despite the great numbers of underemployed agricultural workers, farm machinery is now widespread in the green revolution areas. There are a number of factors promoting this. First, the same companies which supply the agrochemical inputs are keen to increase their machinery sales so they advertise and promote their tractors. But there is also a reason inherent in the social polarization. Landless labourers and the remaining farmers tend to conflict over the price and conditions of labour. Since farming operations such as harvesting need to be done at particular times and with haste this provides a good weapon for industrial action on the part of workers. The mechanization of farm processes undermines the strength of this weapon as well as ensuring that the farm schedule is adhered to (Byres, Crow and Ho, 1983, pp. 29 and 35).

' It should finally be noted that the distribution of the produce is handled through the mechanism of the market. Food is likely to be exported or traded on the international market for manufactured goods, fuel and agrochemicals rather than routed to the poor. While the cultivation of green revolution grains may increase the amount of food a country produces it does not necessarily help to feed that country's poor.

In summary, the case of the green revolution helps to make a number of points. It shows that technical aid can succeed in increasing overall food production but also that such assistance is likely to embody assumptions about the conditions of use which may not be appropriate to the underdeveloped world. The green revolution was a capital–intensive solution to problems of food production in countries where labour is plentiful. It could thus be described as technologically inappropriate. By encouraging reliance on imported seed stock, agrochemicals and farm machinery it can also be said to have enhanced technological dependency. Byres, Crow and Ho (1983, pp. 7–8) note that, at the time the original research was being performed in Mexico, there was another local research project concerned with the improvement of yields through the gradual revision of farmers' customary methods. This did not attract the funding that the technological, green revolution proposal did. It might not have been able to put such funding to good use, but the existence of this other project at least suggests that there might be other ways of directing technical assistance at the problem of food production. Such an alternative strategy to the whole question of technology transfer is embodied in the 'intermediate' or 'alternative' technology movement, to which we now turn.

Intermediate Technology

The tendency for imported technologies to create dual economies is increasingly acknowledged. None the less, since this dual structure has arisen out of an attempt to modernize, it can easily be regarded as the cost which has to be paid in the present for any chance of a modernized future. It is readily regretted but less easily avoided. The prevailing sense among Western academic experts that there was no real alternative to the development of this dual structure was challenged by an economist, Schumacher, in the 1960s and early 1970s. Schumacher was a senior economist with the UK's National Coal Board who was seconded as economic adviser to various Third World countries. He came to argue that an altogether new approach could be adopted to the spread of technology in underdeveloped nations if the investment in technology was deployed differently. His proposal was that a less expensive technology should be devised, pitched somewhere between the high cost of modern, capital–intensive job places and the widespread subsistence economy. This technology should be new and selected for efficacy. But it should also be small scale, relatively simple to understand and capable of location in the rural areas where the majority of the population still lived.

The aim of this technology would in one sense be the same as that pursued by the conventional path of modernization: it should increase output and the level of national wealth. At the same time it would not be divisive since it would be available to far larger numbers of the population. It would not encourage a dual economy nor the social polarization which accompanies it. In the dual economy the people in the modern sector live in urban areas; they consume imported goods; they often receive relatively high levels of education; and they commonly depend on being able to consume the agricultural produce of the rural hinterland. Intermediate technology, being low cost and decentralized, would

167

not have to be restricted to urban areas; it could promote internal trade in the goods people required; it would encourage the spread of education since it would require a uniform but not drastic improvement in people's skills; and it could be geographically integrated with rural production.

To give a simple illustrative idea of his proposals, Schumacher suggested that we think of jobs in the modern sector as £1,000 job places and those in the subsistence sector as £1 job places; an intermediate technology would typically offer £100 job opportunities. These costs (given in the early 1970s and therefore now to be multiplied by a factor of about ten) are intended to illustrate the amount of capital required to furnish each employment opportunity; a workshop in the modern sector providing work for ten people might therefore require an investment of around £10,000. In a country where the dual economy prevails, only a minority of people have the opportunity to work in the modern sector. Schumacher suggests that a typical size for this minority is 15 per cent. Accordingly, for every hundred people in the population the investment in work places can be reckoned as shown in Table 6.1.

Table 6.1 *Illustration of the Cost of Employment Places for Every 100 Members of the Population in the Dual Economy and on Schumacher's Model*

| | Dual economy | | Schumacher's |
	Subsistence	Modern	proposal
Number of people	85	15	100
Average cost (£)	1	1,000	100
Subtotal (£)	85	15,000	10,000
Total (£)		15,085	10,000

What is particularly striking in this very approximate model is that although the intermediate technology is in many respects closer to the cost of the modern sector than to the subsistence technology (the ratios are 1:10 rather than 1:100), on Schumacher's recommendations all the population can have employment opportunities for two–thirds of the current level of investment. Accordingly, with no increase in overall investment in technology, one of the main drawbacks of inappropriate technologies, their tendency to generate unemployment, could be avoided.

Schumacher's arguments were faced with two difficulties: they were unconventional, at least within the Western development tradition, and there were no practical implementations of his ideas to which he could point. Accordingly, the career of his ideas took off in two directions. There was, first, a considerable academic debate about the validity of his ideas; and secondly, organizations began to be founded intent on putting his ideas into practice. By the time Schumacher published his ideas in their best–known form, the book *Small is Beautiful*, the academic debate was well advanced. In this book he provides a summary of what he regards as the leading objections and of the replies to them (1974, pp. 151–5). From this early debate there emerged two principal objections. The first of them is a response to the kind of calculation set out in Table 6.1. The figures in that calculation concerned the cost of creating work opportunities; they did not, however, treat the output to be expected from those jobs. We have seen that in the Western countries labour–saving innovations have gone along with vast increases in the measured productivity of each worker. Thus, while one might accept that Schumacher's scheme would provide more work there is no guarantee that it would produce more goods or more wealth. Jobs in the modern sector which cost, for example, ten times as much to create as jobs founded on intermediate technology could be much more than ten times as productive. And if that were the case, then the underdeveloped country would appear to earn more by investing in the modern sector than by investing in intermediate work places. Schumacher (1974, p. 152) cites this argument in the words of an opponent, N. Kaldor:

If we can employ only a limited number of people in wage labour, then let us employ them in the most productive way, so that they make the biggest possible contribution to the national output, because that will also give the quickest rate of economic growth. You should not go deliberately out of your way to reduce productivity in order to reduce the amount of capital per worker. This seems to me nonsense because you may find that by *increasing capital per worker tenfold you increase the output per worker twentyfold*. There is no question from every point of view of the superiority of the latest and more capitalistic technologies [italics added].

This is certainly an important kind of objection. It is essentially the argument which has become familiar in the West in recent years to the effect that inefficient work places cannot be retained.

Countries have to move to new, more productive technologies, it is argued, even at the cost of very high rates of unemployment. Since this argument has attained such popularity in application to the West it is rather difficult to confront head on; instead one can aim to show that in the context of underdeveloped countries it is misplaced. The argument may not hold for at least three reasons. First, attention has already been drawn to the fact that not all the wealth generated in the modern sector will benefit the host country. Profits may be repatriated or taken away from the host economy through transfer pricing. Even where attempts are made to curb such activities, companies may secure their 'returns on the operation through dividends, profits on the sale of intermediate goods, royalty payments, technical fees or some mixture of these and other forms' (Clark, 1975, p. 15). Not all the output, calculated in Kaldor's terms, can therefore be assumed to benefit the host country.

The second problem with the type of calculation made by Kaldor is that it assumes that the economic situation can be considered in a static manner. It deals only with the output of each work place or set of work places considered in isolation. Within the modern sector there may be some interconnections which mean that enterprise in one area encourages work in a neighbouring or related plant. Such interconnections are likely to be restricted by the pattern of trade with foreign firms in which the modern sector companies engage. In any case the modern sector is likely to have only a small 'knock–on' effect on the subsistence sector. The setting up of intermediate technology businesses on the other hand would, it is argued, encourage such knock–on relationships. By providing work for each other, intermediate enterprises should come to multiply economic activity. In this sense Kaldor's calculation excludes dynamic economic benefits which are likelier with intermediate than with modern technologies.

Kaldor's position also tends to favour modern technology by building into the calculation the assumption that modern technologies will be used at full capacity. Against this Stewart cites a survey suggesting that in the mid–1960s in West Pakistan 'the average industrial firm operated its equipment only one–third of the time' (1978, p. 64). If the use of modern technologies is below capacity their measured efficiency falls rapidly. Since industries in modern sectors are often dependent on imported materials and require high levels of expertise there are many hitches which can disrupt smooth production. The other side of this same coin is that Kaldor assumes that intermediate technologies cannot be made to approach the efficiency of the most modern technologies. This point is at the heart of the second major objection to Schumacher's programme.

The argument is made that an acceptance of an intermediate technology would consign underdeveloped countries forever to a second–rank status. To turn one's back on the modern sector seems to banish altogether the possibility of becoming as wealthy and powerful as the West. This objection is a persuasive one, since the industrial development of the West is easily regarded as the keystone of its success. A twofold response can, however, be made. On the one hand, the whole proposal for an intermediate technology came about because the introduction of modern manufacturing technology did not seem to be having a beneficial effect in many instances. Our study of technological dependency suggests that, in most cases, this situation is not readily altered. Therefore, the objection to intermediate technology is based on false assumptions. If there is no reason to suppose that the modern enclave will grow rapidly enough to benefit the total population in the near future, the advocates of modern technology are also condemning underdeveloped countries to a second–rate status. They will be second–rate because they are modern only in a fraction of their economy and because the capital needed for this part of the economy precludes investment in the remaining sector. One might perhaps argue that being second–rate in this way is somehow psychologically preferable to being distinguished by a policy which avoids modern technologies; one would still feel within reach of Western achievements. It is hard though to see how this could console the poor majority.

The second strand of the reply is more positive. It asserts that adopting intermediate forms of technology need not be a second–rate route at all. This is partly because people will be able to produce goods for each other's use and thus have a knock–on effect encouraging more and more economic activity. More importantly, it is only an *assumption* that intermediate technologies cannot be highly productive. There would seem to be no necessary reason why in many cases a modern technology demanding, say, ten times the investment per work place of an intermediate technology should be more than ten times as productive. In the West technical changes have generally been aimed at substituting labour with inanimate sources of power. Increases in productivity have gone along with a move towards greater capital intensiveness and away from labour intensiveness. For this reason, increases in efficiency have nearly always been associated with greater capital investment per work place. But there has been no systematic incentive to

171

examine ways of increasing productivity whilst keeping other factors equal.

If the image one has of intermediate technology is of a less capital–intensive technology used by the West some time in the past then, as Kaldor claimed, the intermediate technology is (virtually) bound to be less productive than the modern alternative because the modern version actually supplanted the earlier technology. But the modern technology was selected as superior in the context of the whole range of costs prevailing then. If oil was cheap in comparison to the costs of labour then new technologies would tend to be both more productive and labour–replacing. This does not prove that the same old technology could not have been developed in some other way without using inanimate sources of energy to take people's places. As Schumacher notes (1974, p. 156):

> The development of an intermediate technology, therefore, means a genuine forward movement into new territory, where the enormous cost and complication of production methods for the sake of labour saving and job elimination is avoided and technology is made appropriate for labour surplus societies.

Intermediate technology offers an alternative route for improving on old technologies: a route which takes into account the cost of wages, power and raw materials in underdeveloped countries rather than the costs in advanced nations. It should be possible to improve efficiency without having to repeat the path of technical change adopted in the West, and therefore the intermediate technology need not be productively second–rate. The promise is there for an intermediate technology which 'would be immensely more productive than the indigenous technology' (1974, p. 150). The difficulty at the time the book was written was that there were very few actual examples of productive intermediate technologies to which Schumacher could point. Without these examples the promise appeared speculative.

Setting Appropriate Technology to Work

In the intervening years several accounts of intermediate technology have appeared. These relate both to the general issue of the evaluation of Schumacher's ideas and, as mentioned above, to the practical efforts of groups aiming to implement his proposals. The first such group to be established, with Schumacher himself as a founder member, was the Intermediate Technology Development Group (ITDG). A survey of its work has recently been published by one of the co–founders, George McRobie (1982). This book sets out many of the achievements of the ITDG, classified by the type of project undertaken. The headings given are: building; water; farming; transport; energy; health; and women (i.e. technologies to assist women in carrying out their traditional tasks). A brief indication of the nature of the intermediate technologies envisaged by the group can be gained from examining some of the examples offered.

The area of building work provides a useful beginning. McRobie reports (1982, pp. 41–2) that interest centred on two activities: the substitution of innovative local materials for Western products and the scaling down of production methods. The 'modern' standard Portland cement can be replaced by cements based on pozzolanas – naturally ocurring materials produced by certain volcanic processes which react with lime to form an enduring cement. Equally, cement blocks could be better used for construction purposes if they were reinforced. Intermediate technologists investigated the use of natural fibres (vegetable and animal products) for the reinforcement. Bamboo and pine needles have been tried in this way (1982, p. 44). Such reinforced cement can also be formed into shaped components for piping and other uses. The attention focused on production methods has also reportedly paid off. As McRobie states (1982, p. 42):

> Highly efficient small–scale brick and tile units have been developed by the Group's building panel ... These small units produce not one million but 10,000 bricks a week, using hand–operated methods or very simple machinery. Capital costs per work place are about £400 as against £40,000 in a large modern brickworks. In small units, very large savings in fuel costs are possible by air–drying the bricks before firing, and local production virtually eliminates transport costs.

These brief examples display a number of features. There is, first of all, a great reduction in the cost of each employment place. Yet the selection of the intermediate level does not have to bring price disadvantages. Decentralized production reduces transport costs, and the small scale of manufacture allows the use of local materials which might not be handled so readily in a modern industrial plant. Finally, the small volumes of production allow drying to be

173

performed without the use of imported fuel sources and thus make an important saving on a commodity which eats up foreign currency reserves. Intermediate–level production encourages employment without necessarily leading to second–rate or expensive goods.

A second example concerns the provision of drinking water. Dunn (1978, pp. 8–9) describes a method of creating a small–scale rainfall collection reservoir of around 10,000 gallons capacity. An excavated pit is lined with several coats of thin plastic sheeting alternating with mud slurry. A final covering is made up of stacked layers of short tubes of the plastic filled with sand and cement to produce firm, enduring walls. Although this design depends on modern materials, these materials are very inexpensive, and the construction is very labour intensive. Dunn records that the technique was pioneered by ITDG in Botswana. Teachers from rural areas were subsequently invited to the site and taught the construction techniques. They were given a set of the construction materials and were encouraged to involve their pupils in making a reservoir when they returned to their district schools. Around one in four teachers completed such schemes. McRobie suggests that small–scale schemes of this sort form an alternative to expensive projects like drilling deep boreholes. They are cheap and locally based and can readily be understood and constructed by people without special training.

These examples display the promise of intermediate technology projects. They also show that the characteristic focus of many of these projects is on the satisfaction of people's basic needs. Schumacher's own account of intermediate technology had not specified the level of goods at which production would be aimed. Clearly they would not be things available only within the modern sector, but it might have been expected that the goods would themselves be of an intermediate level. In *Small is Beautiful*, Schumacher was unable to supply many illustrative instances, but the examples described above and in more comprehensive accounts (see Carr, 1985) indicate that the intermediate technologies hailed as successes are generally concerned with goods which satisfy basic requirements. There are two principal reasons for this. First, if we take seriously the idea that development with intermediate technology will involve the 85 per cent of people in the subsistence sector, then their needs are going to be very much for basic goods. The second point is that in such cases it is relatively easy to anticipate people's requirements. That the most prominent successes have been recorded at this level leads one to ask what intermediate technology has to offer for subsequent development.

Limitations and Criticisms of Intermediate Technology

The further assessment of intermediate technology can be usefully begun with an example of transport technology cited by McRobie. He justly points out that conventional development approaches to transport have tended to favour road building and related activities which encourage existing transport technologies. As an alternative he cites a boat project based in Southern Sudan. The boatyard constructed fairly large (forty–five foot), barge–like cargo carriers of reinforced cement. Such boats can be built and repaired locally and, because they have a shallow draught, can navigate on the Nile. McRobie (1982, p. 51) states that sales were proceeding well and that 'the project had become self–financing'. In 1985, however, it was reported (Charnock, 1985, p. 10) that this project

> created the best engineering workshop in the country's southern region, training local people in the skills needed to make ferro–cement boats. But the boatyard went out of business. Initial interest in the boats failed to turn into orders. Slow payments created problems, with cash flow, and inexperienced pilots flogged the boats' engines to death. Sudan lost its only boat builder.

The success of the venture was intimately connected not just to assessments of the appropriateness of the technology but also to the marketability of the product. When the technology can be introduced with great confidence in its desirability, as with water provision or inexpensive dwellings, the issue of acceptability is virtually equivalent to technical appropriateness. Where local requirements are more open, the issue is not so cut and dried. This point can be developed in two ways. The first is to note the ways in which intermediate technologists' expectations of people's requirements have sometimes conflicted with people's practical demands. The second avenue examines the nature of the fit between development through intermediate technology and the market mechanism.

The first category of observations has made good journalistic copy under headings such as 'appropriate technology's hall of infamy'. In addition to the case above, Charnock (1985, p. 10) cites one 'classic' gadget: 'the hand–held maize sheller. The gadgets cost next to nothing – but many women can shell quicker with their bare hands.' The technical level may be correct, but that is no guarantee of utility. Equally, Harrison (1983, p. 149) cites the example of moulded concrete latrines which were produced for Orissa in India. These were inexpensive and should have yielded health benefits. Yet they needed comparatively large quantities of water to flush them and so quickly fell into disuse. The point in both these cases is that designers have falsely anticipated people's requirements and have produced goods of no proven utility. Goods must meet people's felt needs and must be construed as working satisfactorily by the people who use them, not by the designers. Designs developed in a Western university may fail either because they are inapplicable in some way or because they cannot stand up to the harsh conditions and repeated use to which they will be subject. If the intended clients are living at a subsistence level it is all the more important that any machinery introduced should work and not leave them worse off than before or committed to a continuing expense they cannot afford. As Harrison comments (1983, p. 149): 'Villagers have a right to expect technical perfection before a new machine is tried out on them at their expense, because they have too little surplus income to waste on failures.' Intermediate technologies, devised by Western engineers, cannot just be assumed to work appropriately.

Other projects have encountered socioeconomic difficulties. For example, schemes to use animal dung more efficiently by converting it into fertilizer slurry and combustible gas have tended to reinforce wealth inequalities since the excrement which was formerly freely available as a source of fuel energy now becomes a valuable commodity. The wealthier people who had the greatest share of the cattle in the first place were in a position virtually to monopolize the dung for gas production and had the best access to capital for funding the converters (Disney, 1977, pp. 96–7). The new technology was satisfactory under test conditions but was by no means uniformly beneficial.

One might be inclined to regard the last case as just a peculiar misfortune. However, it relates closely to the second critical point, concerning the market system into which appropriate technologies are introduced and the kinds of development these techniques are expected to foster. It is clear that Schumacher and his supporters anticipated that people would benefit from appropriate technology through the establishment of small market trading. In the most general sense, people were expected to become small capitalists. But this route to social development may characteristically bring the problems revealed in the case just described. Local inequalities are likely to be magnified by the private acquisition of the new technologies. It has to be remembered that the 85 per cent of people in Schumacher's calculation who are to benefit from the new technology are not a homogeneous mass. As James (1980, p. 72) observes, 'Rural areas ... though typically more equal than the urban sectors of developing countries nonetheless display considerable inequality in the distribution of income.' The possibility that small–scale production and manufacturing opportunities will magnify these existing inequalities is constantly present; it is not just a peculiarity of the gas–production case. Any technology which requires a capital investment will be more readily available to the wealthier people. A technology which encourages more efficient use of unequally distributed resources is likely to amplify those inequalities. It may, of course, be that such is the price to be paid for some improvement in the condition of the poor majority. Equality may have to be traded off for a measure of prosperity. But this is not a price that the original promise of the intermediate technology movement led one to anticipate.

A different line of criticism has also been developed using the case of gas and fertilizer production. Disney sought to draw an economic comparison between the production of fertilizer by intermediate techniques and industrialized techniques. Many parts of the calculation, such as the costs of dung collection and transport, are difficult to assess or, as with the price of industrial fuel, are subject to wide variation. None the less, he concludes that it would be hugely expensive to meet India's fertilizer requirement through decentralized, family–sized production. Such production would be extremely labour intensive – indeed, the high cost is largely due to the labour prices – but the labour would be very inefficiently deployed. The intermediate technology could, however, be otherwise organized, it could be based on village or co–operative units. This arrangement has much higher productivity and could, depending on details of the cost of inputs, rival the highly industrialized techniques (1977, p. 88). Disney reports though that attempts at co–operative organization have not met with much practical success (1977, p. 97), with poor records of supervision and maintenance. Furthermore, the intermediate technique places very heavy demands on the water supply which, for climatic reasons, is impossible to guarantee.

One strand of Disney's critique might not be too disturbing for advocates of appropriate technology; although the rural gas plants have very low productivity it is not clear what else labour could be used to do. However, his other argument – that the ramifications of using intermediate technologies on a national scale are rarely calculated – does suggest limitations to the intermediate technologists' programme.

A further potential difficulty concerns the methods for getting the new techniques to the rural poor. In the example of the water tank described by Dunn the materials were distributed freely to village representatives. This suggests that aid would be the obvious vehicle for introducing appropriate technologies. A number of aid agencies, particularly non–governmental organizations, have begun to endorse this approach (Stewart, 1985, p. xiii). However, we have already seen that aid is dominated by governmental agencies and that their commercial links often incline them to support high–technology transfers. If the success of appropriate technology projects rests on funding from aid agencies the movement will have to look for large changes of policy in the West. In this context it should be noted that the World Bank does have an intermediate technology office, which some see as a promising sign (Harrison, 1985, p. 400). Hayter and Watson though are less sanguine, regarding the office as little more than a token (1985, p. 275).

This brings us to the final criticism commonly levelled at the proposal for intermediate technology, which is that it seeks huge changes in parts of the economy without, in Howes's words, 'prior alteration of other elements in the political, economic and social contexts within which it arises' (1979, p. 117). For example, it calls for a change in patterns of technical investment without examining such issues as landownership or disparities of wealth which will influence the adoption of the new techniques. This issue has already arisen in the context of the fertilizer production where the previously wealthy benefited disproportionately. Equally, little consideration is given to the behaviour of foreign firms and the local élites in the light of attempts to change the technical base; companies which are doing well out of inappropriate technology are hardly likely to endorse the alternative. How, for example, could foreign firms be dissuaded from trying to undercut the new intermediate goods (say, fertilizers) in order to protect their former markets? It will be recalled that Schumacher's original calculation, which suggested the viability of the £100 technology, actually presumed that investment would be diverted away from the £1,000 technologies to finance it. The calculation paid no attention to the political aspects of this switch.

It is not exactly fair to suggest that Schumacher or his followers ignored these issues (Schumacher, 1974, p. 143). But they have placed far less emphasis on political than on technical analyses. In a sense they have walked a policy tightrope, aiming to do something about economic hardship without demanding 'unrealistic' political changes. But critics argue that to go beyond showpiece instances of the new techniques political changes will have to be made and that such changes should really come first in a development strategy. Such critics cite the example of mainland China, where, at least until very recently, measures of economic equality have been enforced with great rigour. Under these circumstances they have achieved widespread use of devices like the bicycle and decentralized medical techniques which might be claimed as intermediate. By limiting the role of the market, restricting people's choice of goods, excluding foreign firms and preventing the build-up of much personal wealth, many inter-mediate measures have been adopted. Of course, the pace and direction of development in China can be questioned. But it can at least be argued that if one's goal is widespread use of intermediate technologies, economic and political regulation is extremely important.

Responses of Intermediate Technologists

Advocates of appropriate technology are currently pursuing a lively debate over policy, and we cannot talk of a single, unified response to the criticisms. Some want to integrate intermediate technologies much more closely into the free market; they wish to engage local and foreign capital in backing designs which appear promising and popular and to get the firms to use their promotional skills to enlarge the market for intermediate goods (Charnock, 1985, p. 11). Supporters of this position are struck by the quality of many intermediate products which they contrast with their low 'market penetration'. Generally expressed, this view could be summarized as a call for better management of intermediate technology. Other supporters look more directly to policy initiatives and to governmental action, citing, for example, the strategy of the Indian government in selectively directing credit to small–scale firms and village industry (Harrison, 1985, p. 401).

In both cases they are recognizing that the social and political context has to be addressed as well as problems of design. But either way they are still confronted by those whose interests may be served by transfers of inappropriate technology and technical dependency.

Concluding Observations

Since we have covered a large number of topics and a great deal of material in this chapter it does not yield a simple conclusion. Two points should be mentioned. First, it should be recalled that the theories of development and dependency which frame the material covered in this chapter are themselves in flux. Dependency theory is partially in retreat; the theorists have been unable to agree about the precise mechanisms by which dependency is sustained and about the classification of degrees of underdevelopment. Equally, the economic, social and (even in some cases) political success of the NICs has thrown doubt on the inescapability of dependency. Technological dependence can therefore no longer be taken as the inevitable outcome of technology transfers. But neither should the success of the NICs be viewed as a simple endorsement of transfers of modern technology. Technology transfer should be studied in detail without presuming success or failure. But as an enduring background we must recall that technology transfers continue to be largely motivated by the benefits accruing to the agency doing the transfer rather than to the recipient.

The second point also concerns an uncertainty. It is very unclear how the problems of international debt are going to be handled. However, countries which face vast repayment charges are extremely unlikely to be able to industrialize rapidly. Debt may lead to the perpetuation of dependence in a new form. How debt and aid are handled will thus be very important in determining whether, and in what form, technology transfers continue and whether support for intermediate technology projects expands.

Notes: Chapter 6

1   Such a view is not the exclusive preserve of advocates of the free market. In most respects it was the expectation of Marx also, a view recently revived by Warren, 1980.

2   It is worth noting that the ATP was introduced under the last Labour government, although it was said to be the price the Cabinet demanded for an increase in aid spending: Tanner, 1984, p. 25.

[7]

Social Construction and Scientific Knowledge

Overview

As described in the Introduction, this text has two chief goals. The first is to review the ways in which science and technology are connected with contemporary social change in the West and in the underdeveloped world. With regard to this goal the text has operated quite straightforwardly as a review. Each chapter has brought together the results of recent research on aspects of this connection and has offered such conclusions as are available. In some cases these 'conclusions' have had to be rather inconclusive: for example, in the case of the effect of science policy on scientists' day–to–day work or the inquiry into the sources of technical innovation. In other cases it has been possible to pass judgement more clearly. This was so, for instance, when we turned to the reasons underlying the mixed fortunes of intermediate technologies and when we examined the continuing role of élite, secretive decision–makers in the choice of military technology. This book has not aimed to describe forthcoming socially influential technologies nor to recommend particular scientific projects which should be adopted by Third–World nations (on this latter issue see the discussion in Clarke, 1985). Instead, the objective in presenting this review is to demonstrate the centrality of science and technology to a broad range of issues of interest to sociologists and to show that the apparently arcane worlds of science and technology repay sociological and political study.

The second goal was of a more theoretical character. In the Introduction, two sociological interpretations of the nature of scientific knowledge were outlined. Both are opposed to the notion that science is a body of neutral expertise which develops largely immune from social influences. The first, political economy view implies that the development of science and technology is continually and profoundly shaped by political and economic considerations. Support for this approach was brought forward in Chapter 4, where the work of analysts like Gorz and Klass was used to indicate that technical innovations are moulded chiefly by the needs of companies and not by a technical logic nor in response to customers' needs.

Social constructionism, the second view, suggested that social influences could always be detected at the heart of scientific and technical judgements. Although these social influences may be of a commercial or overtly political kind, this is far from universally the case. On this view, however, scientific evidence alone can never be sufficient logically to compel people to adopt one particular belief. In principle, the evidence could always be interpreted in alternative ways. The beliefs which people actually do come to hold are thus underdetermined by the evidence from the natural world. Their beliefs are also partly influenced by the social context in which they are assessed. Thus, for example, in Chapter 2 it was argued that the scientific 'constitution' drawn up by early members of the Royal Society was not simply a reflection of how the study of the natural world had to proceed. It outlined a way of studying that world in accordance with particular politico–religious requirements. The demands of natural philosophy underdetermined the construction of the scientific enterprise. Similarly, in Chapter 1, we saw how observation and experimental dissection were essential to the disputes over phrenology but did not uniquely determine the outcome of the dispute nor the beliefs which the protagonists held.

At one level, advocates of the political economy view would be expected to welcome the constructionists' arguments. If social influences are present at the heart of scientific knowledge, it should prove possible to make strong claims about the political economy of science. Indeed, in the case of the phrenological disputes, overtly political issues were addressed by the disputants even when they were engaged in the most minute, technical observations of the brain. However, although a social constructionist analysis was given of the debates over solar neutrinos and gravity waves, nothing obviously commercial or political hinged on the outcome of these controversies. This implies that the political economy view entertains only a restricted notion of the social influences on science. Advocates of this view might reply that vitally important political–economic consequences of scientific and technical products sometimes stem from the effects of those products. How, they might ask, can these effects be analysed if science and technology are treated as mere social constructs?

The discussion in Chapter 5 was concerned with answering this difficulty. Military technology (a technology with a reasonably well defined objective) was examined. Many overtly political and commercial considerations were seen to influence decision-making over military technology; indeed, this was the essence of the claim about the military–industrial complex. However, by screening off occasions when technical decisions were seen to be overridden by such overt considerations, we tried to look at narrowly and (as far as possible) purely technical issues. When we did this, these technical decisions turned out to have a socially negotiated character also. Whether it was the speed of a plane, the utility of a gun, or the firepower of a tank, technical assessments contained a social element. They too, just like the efforts to measure the true flux of neutrinos arriving from the sun, were underdetermined by the evidence available.

No doubt the influence of overtly commercial and political considerations, such as the business opportunities which follow from the choice between competing fighter designs, commonly pre–empts technical decision making. No doubt civil technologies also are pervasively shaped in this way. That this occurs does not demonstrate that, under other circumstances, purely technical decisions could be made. No scientific or technical decision is free of elements of social construction.

Moderate Constructionism

In asserting this position I have no wish to imply that observation and testing, careful trial and measurement are absent from scientific and technical choice. Sometimes, perhaps for political reasons, they may be absent but characteristically they are not. They were not absent from the decision–making over Nimrod nor from the dispute over phrenology. Rather, the argument being proposed here is that, when an issue is subject to technical decision–making, the outcome will be underdetermined by the evidence at hand. In principle, alternative interpretations of the outcome are always possible. Social negotiations will complement the underdetermining technical evidence.

On many occasions the social component may, to common sense, appear slight. It may simply be a matter of the trust and respect which the parties hold for each other's experimental skill and honesty. But assessments of skill and honesty in their turn may well be predicated on less personal, more sociological, or even political evaluations. Thus, in the case of phrenology, one of the dimensions along which people's testimony was assessed was their status as amateur or authorized medical practitioners. In the dispute over the technical merits of SDI, the question of whether people are 'hawks' or 'doves' routinely plays a part in the reception given to the arguments and evidence they present. And in the debate over Nimrod, the party political allegiance and armed–service or company affiliation of speakers was used by listeners in weighing up the value of their claims. No firm border-line can be drawn around the kinds of social issues which will be used by people in the course of technical decision–making. No sharp distinction can separate 'internal' and 'external' factors. When, therefore, sociologists study areas of social life in which technical decisions play a part, they should be alert to the elements of social construction which inform those decisions.

The characteristic concern of the political economy approach with overtly commercial and political influences can often direct analysts to issues of enormous practical significance. But only a thorough social construction approach, which recognizes how the apparently 'external' fuses with the apparently 'internal', can indicate how fully social are decisions based on scientific and technical reasoning.

The position advocated here might be termed 'moderate constructionism'. Science and technology are not *mere* social constructions; but constructions they are all the same. In Chapter 1 we examined arguments for the rationality of science put forward by Newton–Smith. He concludes his argument by calling for a 'temperate rationalism' (1981, pp. 266–73). Loosely expressed, he wishes to argue that scientific knowledge is more of a construct than most philosophers have been willing to admit but that it is a very special construct which tends to become truer and truer. Although I share his moderating language and am trying to stake out a position between epistemological extremes, real distinctions between us remain. We are not just disputing whether a beer glass is half empty or half full.

For example, in accounting for the historical success of science as a form of knowledge, Newton–Smith would point principally to its cognitive merits and technical utility. When I addressed this issue in Chapter 2, I drew attention also to the rhetorical and institutional resources which scientists had mobilized and to science's ideological usefulness which had attracted political sponsorship for it. Similarly, when studying technical change, rationalists – even temperate ones – would explain change in terms of technical improvement while constructionists would focus on the negotiations which underlie the decision that, for example, the new fighter counts as an improvement. The moderate constructionist does not wish to do science down, but wishes to point out that there is inevitably an element of social construction in all decision–making. That element may have great political and economic consequences.

We can conclude with an illustration of this last point. In disputes about arcane technical matters (say, about neutrinos) the element of social negotiation and construction will probably not be of public concern. But scientists and technologists are increasingly called on to make decisions which will affect the public, its health, diet, or environment. The rationalist view of decison-making tends to be too sanguine here. Often an issue is of such urgency, as with public health questions arising from AIDS (Nelkin and Hilgartner, 1986), that it comes to public attention before scientists have arrived at a decision. The constructionist's advice to look for the impact of social and, perhaps, political influences on the apparently 'internal' scientific and medical issues surrounding the case is clearly of practical value in such instances.

Equally, there are some cases, such as the siting of dumps for hazardous waste or the evaluation of risks from radiation, where decisions plainly have political as well as technical aspects. If the decision is disputed, as it commonly is, an attractive strategy for the disputants is to try to claim that their opponents' supposedly factual statements are contaminated by political considerations. Their own technical arguments, on the other hand, are presented as purely factual (Wynne, 1982, pp. 163–5). Both sides frequently adopt this tactic. Called on to analyse this situation, the rationalist is likely to say that, in the long run, one or other will be shown to be correct. The other will have been in error. If anything, this diagnosis will tend to fuel both sides' feelings of self–righteousness. The constructionist, on the other hand, can describe the situation in a more even–handed way. He or she can accept that both sides may have 'internally' technical arguments. The technical arguments may none the less diverge since, as well as evidence from the natural world, they will embody social and political elements. Thus, the constructionist view allows us to explain how a technical dispute can continue unresolved (often growing increasingly acrimonious as accusations of political bias are levelled). It prepares us for the frequently contentious nature of science in public and does not tend to exacerbate such disputes.

The constructionist clearly accepts that science and technology are the best resources we have for dealing with the natural world. But their pre–eminence should not insulate them from critical scrutiny. Constructionism indicates how that scrutiny can best be channelled. And by showing that the evidence underdetermines scientific and technical decisions, this view allows us to explain why, particularly in matters of public controversy, scientific experts often cannot deliver the definitive answers which a simple, scientistic faith often leads planners, politicians, protesters and (sometimes) even scientists themselves to expect.

References

Abir-Am, P. (1982), 'The discourse of physical power and biological knowledge in the 1930s: a reappraisal of the Rockefeller Foundation's "policy" in molecular biology', *Social Studies of Science*, vol. 12, pp. 341-82.

Adejugbe, M. (1984), 'The myths and realities of Nigeria's business indigenization', *Development and Change*, vol. 15, pp. 577-92.

Albury, D., and Schwartz, J. (1982), *Partial Progress: The Politics of Science and Technology* (London: Pluto).

Baran, P. A., and Sweezy, P. M. (1968), *Monopoly Capital* (Harmondsworth: Penguin).

Barnes, B. (1977), *Interests and the Growth of Knowledge* (London: Routledge & Kegan Paul).

Barnes, S. B. (1971), 'Making out in industrial research', *Science Studies*, vol. 1, pp. 157-75.

Basalla, G. (1967), 'The spread of Western science', *Science*, vol. 156, 5 May, pp. 677-704.

Bell, C. (1973), 'The acquisition of agricultural technology: its determinants and effects', in C. Cooper (ed.), *Science, Technology and Development* (London: Frank Cass), pp. 123-59.

Berger, P. L. (1987), *The Capitalist Revolution* (Aldershot: Gower).

Berger, P. L., and Luckmann, T. (1971), *The Social Construction of Reality* (Harmondsworth: Penguin).

Berghahn, V. R. (1981), *Militarism* (Cambridge: Cambridge University Press).

Berman, M. (1978), *Social Change and Scientific Organization: The Royal Institution, 1799-1844* (London: Heinemann).

Bianchi-Streit, M., Blackburne, N., Budde, R., Reitz, H., Sagnell, B., Schmied, H., and Schorr, B. (1984), *Economic Utility Resulting from CERN Contracts* (Geneva: CERN).

Bienefeld, M. (1981), 'Dependency and the newly industrialising countries', in D. Seers (ed.), *Dependency Theory* (London: Frances Pinter), pp. 79-96.

Bowler, P. J. (1984), *Evolution: The History of an Idea* (Berkeley, Calif.: University of California Press).

Broad, W. (1985), 'The scientists of Star Wars', *Granta*, vol. 16, pp. 81-106.

Byres, T. J., Crow, B., and Ho, M. W. (1983), *The Green Revolution in India* (Milton Keynes: Open University Press).

Cannon, S. F. (1978), *Science in Culture: The Early Victorian Period* (New York: Dawson/Science History Publications).

Cardwell, D. S. L. (1972a), *The Organisation of Science in England* (London: Heinemann).

Cardwell, D. S. L. (1972b), *Technology, Science and History* (London: Heinemann).

Carr, M. (ed.) (1985), *The AT Reader: Theory and Practice in Appropriate Technology* (London: Intermediate Technology Publications).

Central Statistical Office (1987), *Social Trends*, vol. 17 (London: HMSO).

Chapman, I. D., and Farina, C. (1983), 'Peer review and national need', *Research Policy*, vol. 12, pp. 317-27.

Charnock, A. (1985), 'Appropriate technology goes to market', *New Scientist*, vol. 106, 9 May, pp. 10-11.

Choudhuri, A. R. (1985), 'Practising Western science outside the West', *Social Studies of Science*, vol. 15, pp. 475-505.

Chubin, D. E. (1984), 'Ethnography of kept science', *European Association for the Study of Science and Technology (EASST) Newsletter*, vol. 3, pp. 5-9.

Clark, N. (1975), 'The multi-national corporation: the transfer of technology and dependence', *Development and Change*, vol. 6, pp. 5-21.

Clark, N. (1980), 'The economic behaviour of research institutions in developing countries', *Social Studies of Science*, vol. 10, pp. 75-93.

Clarke, R. (1985), *Science and Technology in World Development* (Oxford: Oxford University Press).

Collins, H. M. (1975), 'The seven sexes: a study in the sociology of a phenomenon, or the replication of experiments in physics', *Sociology*, vol. 9, pp. 205-24.

Collins, H. M. (ed.) (1981), *Knowledge and Controversy: Studies of Modern Natural Science*, special issue of *Social Studies of Science*, vol. 11, no. 1.

Collins, H. M. (1983a), 'An empirical relativist programme in the sociology of scientific knowledge', in K. D. Knorr-Cetina and M. Mulkay (eds), *Science Observed* (London: Sage), pp. 85-113.

Collins, H. M. (1983b), 'The sociology of scientific knowledge: studies of contemporary science', *Annual Review of Sociology*, vol. 9, pp. 265-85.

Collins, H. M. (1983c), 'Scientific knowledge and science policy: some foreseeable implications', *European Association for the Study of Science and Technology (EASST) Newsletter*, vol. 2, pp. 5-8.

Collins, H. M. (1985a), *Changing Order* (London: Sage).

Collins, H. M. (1985b), 'The possibilities of science policy', *Social Studies of Science*, vol. 15, pp. 554-8.

Cooper, C. (1973), 'Science, technology and production in the under-developed countries: an introduction', in C. Cooper (ed.), *Science, Technology and Development* (London: Frank Cass), pp. 1-18.

Cowan, R. S. (1985), 'How the refrigerator got its hum', in MacKenzie and Wajcman, op. cit., pp. 202-18.

References

Department of Education and Science (1968), *The Proposed 300 GeV Accelerator*, Cmnd 3503 (London: HMSO).

Department of Education and Science (1985), *High Energy Particle Physics in the United Kingdom* (Stanmore, Middlesex: DES).

Department of Industry and Commerce (1966), *Science and Irish Economic Development*, 2 vols. (Dublin: Republic of Ireland Stationery Office).

Disney, R. (1977), 'Economics of "Gobar-Gas" versus fertilizer: a critique of intermediate technology', *Development and Change*, vol. 8, pp. 77-102.

Dunn, P. D. (1978), *Appropriate Technology* (London: Macmillan).

The Economist (1984), *The World in Figures* (London: Economist Publications).

Ellis, N. D. (1972), 'The occupation of science', in B. Barnes (ed.), *Sociology of Science* (Harmondsworth: Penguin), pp. 188-205.

Erickson, J. (1971), 'The military-industrial complex', *Science Studies*, vol. 1, pp. 225-33.

Fallows, J. (1985), 'The American army and the M-16 rifle', in MacKenzie and Wajcman, op. cit., pp. 239-51.

Farina, C., and Gibbons, M. (1979), 'A quantitative analysis of the Science Research Council's policy of "selectivity and concentration"', *Research Policy*, vol. 8, pp. 306-38.

Farina, C., and Gibbons, M. (1981), 'The impact of the Science Research Council's policy of selectivity and concentration on average levels of research support: 1965-1974', *Research Policy*, vol. 10, pp. 202-20.

Fitzgerald, D. (1986), 'Exporting American agriculture: the Rockefeller Foundation in Mexico, 1943-53', *Social Studies of Science*, vol. 16, pp. 457-83.

Frame, J. D. (1979), 'National economic resources and the production of research in lesser developed countries', *Social Studies of Science*, vol. 9, pp. 233-46.

Fransman, M. (1986), *Technology and Economic Development* (Brighton: Wheatsheaf).

Freeman, C. (1974), *The Economics of Industrial Innovation* (Harmondsworth: Penguin).

Galbraith, J. K. (1974), *The New Industrial State* (Harmondsworth: Penguin).

George, S. (1977), *How the Other Half Dies* (Harmondsworth: Penguin).

Gibbons, M. (1970), 'The CERN 300 GeV accelerator: a case study in the application of the Weinberg criteria', *Minerva*, vol. 8, pp. 180-91.

Gibbons, M., and Johnston, R. (1974), 'The roles of science in technological innovation', *Research Policy*, vol. 3, pp. 220-42.

Giddens, A. (1979), *Central Problems in Social Theory* (London: Macmillan).

Gieryn, T. F. (1983), 'Boundary work and the demarcation of science from non-science: strains and interests in professional ideologies of scientists', *American Sociological Review*, vol. 48, pp. 781-95.

Gilbert, G. N., and Mulkay, M. (1984), *Opening Pandora's Box: A Sociological Analysis of Scientists' Discourse* (Cambridge: Cambridge University Press).

Goonatilake, S. (1984), *Aborted Discovery: Science and Creativity in the Third World* (London: Zed).

Gorz, A. (1976), 'Technology, technicians and class struggle', in A. Gorz (ed.), *The Division of Labour* (Brighton: Harvester), pp. 159-89.

Gorz, A. (1980), 'The scientist as worker', in R. Arditti, P. Brennan and S. Cavrak (eds.), *Science and Liberation* (Boston, Mass.: South End), pp. 267-79.

Graham, L. R. (1985), 'The socio-political roots of Boris Hessen: Soviet Marxism and the history of science', *Social Studies of Science*, vol. 15, pp. 705-22.

Gummett, P. (1984), 'Defence research policy', in M. Goldsmith (ed.), *UK Science Policy* (Harlow: Longman), pp. 57-81.

Gummett, P. (1986), 'What price military research?', *New Scientist*, vol. 110, 19 June, pp. 60-3.

Habermas, J. (1971), *Toward a Rational Society* (London: Heinemann).

Habermas, J. (1974), 'Habermas talking: an interview', *Theory and Society*, vol. 1, pp. 37-58.

Habermas, J. (1976), *Legitimation Crisis* (London: Heinemann).

Hales, M. (1982), *Science or Society? The Politics of the Work of Scientists* (London: Pan).

Hanson, N. R. (1965), *Patterns of Discovery* (Cambridge: Cambridge University Press).

Harrison, P. (1983), *The Third World Tomorrow* (Harmondsworth: Penguin).

Harrison, P. (1985), 'Small is appropriate', in Carr, op. cit., pp. 399-403.

Hay, C. (1987), 'The finance of science in the past: patronage in early modern Europe', paper given at the annual meeting of the British Sociological Association, April.

Hayter, T., and Watson, C. (1985), *Aid: Rhetoric and Reality* (London: Pluto).

Herrera, A. (1973), 'Social determinants of science in Latin America', in C. Cooper (ed.), *Science, Technology and Development* (London: Frank Cass), pp. 19-37.

Hessen, B. (1931), 'The social and economic roots of Newton's *Principia*', in N. I. Bukharin *et al* (eds.), *Science at the Crossroads* (London: Kniga), part 9, pp. 1-62.

Hoogvelt, A. (1980), 'Indigenization and technological dependency', *Development and Change*, vol. 11, pp. 257-72.

Hoogvelt, A. (1982), *The Third World in Global Development* (London: Macmillan).

House of Commons (1982), *Appropriation Accounts 1981–82*, vol. 7, class X (London: HMSO).

House of Commons (1984), *Appropriation Accounts 1983–84*, vol. 7, class X (London: HMSO).

House of Commons (1985), *Appropriation Accounts 1984–85*, vol. 7, class X (London: HMSO).

House of Commons (1986a), *Hansard*, sixth series, vol. 93, 'Written answers'.

House of Commons (1986b), *Hansard*, sixth series, vol. 107.

Howe, R. W. (1981), *Weapons* (London: Sphere).

Howes, M. (1979), 'Appropriate technology: a critical evaluation of the concept and the movement', *Development and Change*, vol. 10, pp. 115-24.

Illinois Institute of Technology (IIT) (1968), *Technology in Retrospect and Critical Events in Science* (Chicago: IIT).

Irvine, J., and Martin, B. R. (1984a), 'What direction for basic scientific research?', in M. Gibbons, P. Gummett and B. M. Udgaonkar (eds.), *Science and Technology Policy in the 1980s and Beyond* (Harlow: Longman), pp. 67-98.

Irvine, J., and Martin, B. R. (1984b), *Foresight in Science* (London: Frances Pinter).

Irvine, J., and Martin, B. R. (1985), 'Basic research in the East and West: a comparison of the scientific performance of high-energy physics accelerators', *Social Studies of Science*, vol. 15, pp. 293-341.

Jacob, J. R. (1975), 'Restoration, reformation and the origins of the Royal Society', *History of Science*, vol. 13, pp. 155-76.

James, J. (1980), 'Appropriate technologies and inappropriate policy instruments', *Development and Change*, vol. 11, pp. 65-76.

Kaldor, M. (1983), *The Baroque Arsenal* (London: Sphere).

Klass, A. (1975), *There's Gold in Them Thar Pills* (Harmondsworth: Penguin).

Kornhauser, W. (1962), *Scientists in Industry: Conflict and Accommodation* (Berkeley, Calif.: University of California Press).

Krige, J., and Pestre, D. (1985), 'A critique of Irvine and Martin's methodology for evaluating big science', *Social Studies of Science*, vol. 15, pp. 525-39.

Kuhn, T. S. (1970), *The Structure of Scientific Revolutions*, 2nd edn (Chicago: University of Chicago Press).

Kuhn, T. S. (1977), *The Essential Tension* (Chicago: University of Chicago Press).

Lakatos, I. (1978), *Mathematics, Science and Epistemology. Philosophical Papers*, Vol. 2 (Cambridge: Cambridge University Press).

Lever, H.; and Huhne, C. (1985), *Debt and Danger* (Harmondsworth: Penguin).

MacKenzie, D. (1978), 'Statistical theory and social interests: a case study', *Social Studies of Science*, vol. 8, pp. 35-83.

MacKenzie, D. (1983), 'Militarism and socialist theory', *Capital and Class*, vol. 19, pp. 33-73.

MacKenzie, D., and Wajcman, J. (1985a), 'Introduction on military technology', in MacKenzie and Wajcman, op. cit., pp. 224-32.

MacKenzie, D., and Wajcman, J. (eds.) (1985b), *The Social Shaping of Technology* (Milton Keynes: Open University Press).

Mackintosh, D. (1866), 'Comparative anthropology of England and Wales', *Anthropological Review*, vol. 4, pp. 1-21.

Mangold, T. (1987), 'Is "Star Wars" built on bad science and false promises?', *The Listener*, vol. 117, 15 January, pp. 4-5 and 8.

Marx, K. (1971 [1894]), *Das Kapital*, Vol. 3 (Frankfurt: Ullstein).

McNamara, R. (1987), *Blundering into Disaster* (London: Bloomsbury).

McRobic, G. (1982), *Small is Possible* (London: Sphere).

Mendelsohn, E. (1977), 'The social construction of scientific knowledge', in E. Mendelsohn, P. Weingart and R. Whitley (eds.), *The Social Production of Scientific Knowledge, Sociology of the Sciences Yearbook 1* (Dordrecht: Reidel), pp. 3-26.

Merton, R. K. (1973), *The Sociology of Science* (Chicago: University of Chicago Press).

Milne, A. (1984), 'The high-tech military and civilian alliance', *The Listener*, vol. 111, 10 May, pp. 12-13.

Mowery, D. C., and Rosenberg, N. (1982), 'The influence of market demand upon innovation: a critical review of some recent empirical studies', in Rosenberg, op. cit., pp. 193-241.

Mulkay, M. (1976), 'Norms and ideology in science', *Social Science Information*, vol. 15, pp. 637-56.

Mulkay, M. (1980), 'Sociology of science in the West', *Current Sociology*, vol. 28, pp. 1-184.

Mulligan, L. and Mulligan, E. (1981), 'Reconstructing Restoration science: styles of leadership and social composition of the early Royal Society', *Social Studies of Science*, vol. 11, pp. 327-64.

Myers, G. (1985), 'The social construction of two biologists' proposals', *Written Communication*, vol. 2, pp. 219-45.

NBST (1982), *Annual Report 1981* (Dublin: NBST).

NBST (1983), *Science Budget 1983* (Dublin: NBST).

Nelkin, D., and Hilgartner, S. (1986), 'Disputed dimensions of risk: a public school controversy over AIDS', *Milbank Quarterly*, vol. 64, pp. 118-42.

Nelson, D. (1973), 'The support of university research by the Science Research Council', *British Journal of Political Science*, vol. 3, pp. 113-28.

Newton-Smith, W. H. (1981), *The Rationality of Science* (London: Routledge & Kegan Paul).

OECD (1974), *Reviews of National Science Policy: Ireland* (Paris: OECD).

Ozanne, R. (1967), *A Century of Labor-Management Relations at McCormick and International Harvester* (Madison, Wis.: University of Wisconsin Press).

Palma, G. (1981), 'Dependency and development: a critical overview', in D. Seers (ed.), *Dependency Theory* (London: Frances Pinter), pp. 20-78.

Pavitt, K., and Worboys, M. (1977), *Science, Technology and the Modern Industrial State* (London: Butterworths).

Pearse, A. (1980), *Seeds of Plenty, Seeds of Want* (Oxford: Clarendon).

Pinch, T. J. (1981), 'The sun-set: the presentation of certainty in scientific life', *Social Studies of Science*, vol. 11, pp. 131-58.

Porter, R. (1973), 'The Industrial Revolution and the rise of the science of geology', in M. Teich and R. Young (eds.), *Changing Perspectives in the History of Science* (London: Heinemann), pp. 320-43.

Potter, J., and Mulkay, M. (1982), 'Making theory useful: utility account-ing in social psychologists' discourse', *Fundamenta Scientiae*, vol. 3, pp. 259-78.

Price, D. J. de Solla (1965), *Little Science, Big Science* (New York: Columbia University Press).

Proxmire, W. (1970), *Report from Wasteland* (London: Praeger).

Reid, W. (1899), *Memoirs and Correspondence of Lyon Playfair* (London: Harper & Bros).

Rip, A. (1982), 'De gans met de gouden eieren en andere maatschappelijke legitimaties van de moderne wetenschap', *De Gids*, vol. 145, pp. 285-97.

Ronayne, J. (1984), *Science in Government* (London: Edward Arnold).

Rose, H., and Rose, S. (1970), *Science and Society* (Harmondsworth: Penguin).

Rosenberg, N. (1982), *Inside the Black Box: Technology and Economics* (Cambridge: Cambridge University Press).

Roslender, R. (1987), 'Toward a new perspective for the study of scientific workers', paper given at the annual meeting of the British Sociological Association, April.

Rostow, W. W. (1971), *The Stages of Economic Growth* (Cambridge: Cambridge University Press).

Roxborough, I. (1979), *Theories of Underdevelopment* (London: Macmillan).

Rudwick, M. J. S. (1963), 'The foundation of the Geological Society of London: its scheme for co-operative research and its struggle for independence', *British Journal for the History of Science*, vol. 1, pp. 325-55.

Russell, C. (1983), *Science and Social Change: 1700-1900* (London: Macmillan).

Salomon, J.-J. (1973), *Science and Politics* (London: Macmillan).

Schmied, H. (1977), 'A study of economic utility resulting from CERN contracts', *IEEE Transactions in Engineering Management*, vol. 24, pp. 125-38.

Schumacher, E. F. (1974), *Small is Beautiful*, 2nd edn (London: Sphere).

Science and Engineering Research Council (1982), *Report of the Science and Engineering Research Council for the Year 1981-82* (Swindon: SERC).

Science and Engineering Research Council (1983), *Report of the Science and Engineering Research Council for the Year 1982-83* (Swindon: SERC).

Science and Engineering Research Council (1984), *Report of the Science and Engineering Research Council for the Year 1983-84* (Swindon: SERC).

Science and Engineering Research Council (1985), *Report of the Science and Engineering Research Council for the Year 1984-85* (Swindon: SERC).

Sensat, J. (1979), *Habermas and Marxism: An Appraisal* (London: Sage).

Shapin, S. (1979), 'The politics of observation: cerebral anatomy and social interests in the Edinburgh phrenology disputes', in R. Wallis (ed.), *On the Margins of Science, Sociological Review Monograph* Vol. 27 (Keele: Keele University Press), pp. 139-78.

Shapin, S. (1982), 'History of science and its sociological reconstructions', *History of Science*, vol. 20, pp. 157-211.

Sherwin, C. W., and Isenson, R. S. (1967), 'Project Hindsight: a Defense Department study of the utility of research', *Science*, vol. 156, 23 June, pp. 1571-7.
Slack, J. (1972), 'Class struggle among the molecules', in T. Pateman (ed.), *Countercourse* (Harmondsworth: Penguin), pp. 202-17.
Smith, D., and Smith, R. (1983), *The Economics of Militarism* (London: Pluto).
Smith, M., McLoughlin, J., Large, P., and Chapman, R. (1985), *Asia's New Industrial World* (London: Methuen).
Stewart, F. (1978), *Technology and Underdevelopment*, 2nd edn (London: Macmillan).
Stewart, F. (1985), 'Introduction', in Carr, op. cit., pp. xiii-xvi.
Syfret, R. H. (1950), 'Some early critics of the Royal Society', *Notes and Records of the Royal Society*, vol. 8, pp. 20-64.
Tanner, J. (1984), 'Business strings tie up Britain's programme of overseas aid', *The Guardian*, 7 November, p. 25.
Thompson, B. (1985), 'What is Star Wars?', in E. P. Thompson, op. cit., pp. 28-49.
Thompson, E. P. (ed.) (1985a), *Star Wars* (Harmondsworth: Penguin).
Thompson, E. P. (1985b), 'Why is Star Wars?', in E. P. Thompson, op. cit., pp. 9-27.
Tobey, R. C. (1971), *The American Ideology of National Science* (Pittsburgh, Pa: University of Pittsburgh Press).
Toulmin, S. E. (1961), *Foresight and Understanding* (London: Hutchinson).
Turner, J., and SIPRI (1985), *Arms in the '80s* (London: Taylor & Francis).
Turney, J. (1983), 'Is big really beautiful?', *Times Higher Education Supplement*, vol. 580, 16 December, p. 6.
Tyndall, J. (1905), *Fragments of Science*, Part I (New York: Collier).
UN (1986), *1983/84 Statistical Yearbook* (New York: UN).
UNESCO (1986), *Statistical Yearbook 1986* (Paris: UNESCO).
van den Daele, W. (1977), 'The social construction of science', in E. Mendelsohn, P. Weingart and R. Whitley (eds.), *The Social Production of Scientific Knowledge, Sociology of the Sciences Yearbook 1* (Dordrecht: Reidel), pp. 27-54.
Vessuri, H. M. C. (1986), 'The universities, scientific research and the national interest in Latin America', *Minerva*, vol. 24, pp. 1-38.
Warren, B. (1980), *Imperialism, Pioneer of Capitalism* (London: New Left Books).
Webster, A. (1984), *Introduction to the Sociology of Development* (London: Macmillan).
Weinberg, A. (1963), 'Criteria for scientific choice', *Minerva*, vol. 1, pp. 159-71.
Whitley, R. (1978), 'Types of science, organizational strategies and patterns of work in research laboratories in different scientific fields', *Social Science Information*, vol. 17, pp. 427-47.
Williams, B. R. (1964), 'Research and economic growth - what should we expect?', *Minerva*, vol. 3, pp. 57-71.

Williams, R. J. P. (1986), 'The corridors of cash', *Chemistry in Britain*, vol. 22, pp. 307-8.

Winner, L. (1985), 'Do artifacts have politics?', in MacKenzie and Wajcman, op. cit., pp. 26-38.

Wood, P. B. (1980), 'Methodology and apologetics: Thomas Sprat's *History of the Royal Society*', *British Journal for the History of Science*, vol. 13, pp. 1-26.

Wynne, B. (1982), *Rationality and Ritual* (Chalfont St Giles: British Society for the History of Science).

Yearley, S. (1984a), 'Proofs and reputations: Sir James Hall and the use of classification devices in scientific argument', *Earth Sciences History*, vol. 3, pp. 25-43.

Yearley, S. (1984b), *Science and Sociological Practice* (Milton Keynes: Open University Press).

Yearley, S. (1984c), 'On the argumentative strategies of scientists in the public realm', *Zeitschrift für Wissenschaftsforschung*, vol. 3, pp. 29-37.

Yearley, S. (1985), 'Representing geology: textual structures in the pedagogical presentation of science', in T. Shinn and R. Whitley (eds.), *Expository Science: Forms and Functions of Popularisation, Sociology of the Sciences Yearbook 9* (Dordrecht: Reidel), pp. 79-101.

Yearley, S. (1987a), 'Precambrian studies as interdisciplinary science', *Philosophy and Social Action*, vol. 13, pp. 79-89.

Yearley, S. (1987b), 'The social construction of national scientific profiles: a case study of the Irish Republic' *Social Science Information*, vol. 26, pp. 191-210.

Ziman, J. M. (1981), 'What are the options? Social determinants of personal research plans', *Minerva*, vol. 19, pp. 1-42.

Ziman, J. M. (1983), 'The collectivization of science', *Proceedings of the Royal Society of London*, B. vol. 219, pp 1-19.

Index